# INTERNATIONAL PUBLIC ADMINISTRATIONS IN ENVIRONMENTAL GOVERNANCE

Combining theoretical and empirical approaches, this book examines the role that international public administrations play in global environmental politics in the Anthropocene. With chapters written by leading experts in the field, the book offers fresh insight into how international bureaucracies shape global policies in the complex areas of climate change, biodiversity, and development policy. International public administrations are recognized as partially autonomous actors with their own interests and motivations, assuming the roles of managers, orchestrators, brokers, or attention-seekers. This comprehensive resource provides scholars and practitioners with valuable insight into environmental policymaking and how international public administrations might be transformed to better address the multiple, fundamental challenges of our century. This is one of a series of publications associated with the Earth System Governance Project. For more publications, see www.cambridge.org/earth-system-governance. This title is also available as Open Access on Cambridge Core.

HELGE JÖRGENS is Associate Professor at the Department of Political Science and Public Policy at Iscte – University Institute of Lisbon and an integrated researcher at CIES-IUL, the Centre for Research and Studies in Sociology, Lisbon. He has coedited several books on environmental policy, including *Understanding Environmental Policy Convergence* (Cambridge University Press, 2014), *A Guide to EU Renewable Energy Policy* (Edward Elgar, 2017), and the *Routledge Handbook of Environmental Policy* (Routledge, 2023).

NINA KOLLECK is Full Professor at the University of Potsdam, Germany. Prior to her current role, she held professorships at Leipzig University, RWTH Aachen University, and the Free University Berlin. Additionally, she has served as a visiting Professor at institutions such as the University of California, Berkeley, the University of British Columbia in Vancouver, and the Hebrew University of Jerusalem.

MAREIKE WELL is a Ph.D. candidate at the Free University Berlin. She works at the German Federal Foreign Office on international and European climate policy and climate diplomacy. Her research interests include multilateral negotiations on climate change, biodiversity, and educational issues. Her publications have appeared in *Global Governance*, *Global Environmental Politics*, and *Climate Policy*.

The **Earth System Governance Project** was established in 2009 as a core project of the International Human Dimensions Programme on Global Environmental Change. Since then, the Project has evolved into the largest social science research network in the area of sustainability and governance. The Earth System Governance Project explores political solutions and novel, more effective governance mechanisms to cope with the current transitions in the socio-ecological systems of our planet. The normative context of this research is sustainable development; earth system governance is not only a question of institutional effectiveness, but also of political legitimacy and social justice.

The **Earth System Governance series** with Cambridge University Press publishes the main research findings and synthesis volumes from the Project's first ten years of operation.

**Series Editor**

Frank Biermann, Utrecht University, the Netherlands

**Titles in print in this series**

Biermann and Lövbrand (eds.), *Anthropocene Encounters: New Directions in Green Political Thinking*

van der Heijden, Bulkeley, and Certomà (eds.), *Urban Climate Politics: Agency and Empowerment*

Linnér and Wibeck, *Sustainability Transformations: Agents and Drivers across Societies*

Betsill, Benney, and Gerlak (eds.), *Agency in Earth System Governance*

Biermann and Kim (eds.), *Architectures of Earth System Governance: Institutional Complexity and Structural Transformation*

Baber and Bartlett (eds.), *Democratic Norms of Earth System Governance*

Djalante and Siebenhüner (eds.), *Adaptiveness: Changing Earth System Governance*

Behrman and Kent (eds.), *Climate Refugees*

Lamalle and Stoett (eds.), *Representations and Rights of the Environment*

Biermann, Hickmann, and Sénit (eds.), *The Political Impact of the Sustainable Development Goals: Transforming Governance through Global Goals?*

# INTERNATIONAL PUBLIC ADMINISTRATIONS IN ENVIRONMENTAL GOVERNANCE

The Role of Autonomy, Agency, and the Quest for Attention

*Edited by*

**HELGE JÖRGENS**
*Iscte – University Institute of Lisbon*

**NINA KOLLECK**
*University of Potsdam*

**MAREIKE WELL**
*Free University Berlin*

Shaftesbury Road, Cambridge CB2 8EA, United Kingdom

One Liberty Plaza, 20th Floor, New York, NY 10006, USA

477 Williamstown Road, Port Melbourne, VIC 3207, Australia

314–321, 3rd Floor, Plot 3, Splendor Forum, Jasola District Centre, New Delhi – 110025, India

103 Penang Road, #05–06/07, Visioncrest Commercial, Singapore 238467

Cambridge University Press is part of Cambridge University Press & Assessment, a department of the University of Cambridge.

We share the University's mission to contribute to society through the pursuit of education, learning and research at the highest international levels of excellence.

www.cambridge.org
Information on this title: www.cambridge.org/9781009383462
DOI: 10.1017/9781009383486

© Cambridge University Press & Assessment 2024

This work is in copyright. It is subject to statutory exceptions and to the provisions of relevant licensing agreements; with the exception of the Creative Commons version the link for which is provided below, no reproduction of any part of this work may take place without the written permission of Cambridge University Press.

An online version of this work is published at doi.org/10.1017/9781009383486 under a Creative Commons Open Access license CC-BY-NC-ND 4.0 which permits re-use, distribution and reproduction in any medium for non-commercial purposes providing appropriate credit to the original work is given. You may not distribute derivative works without permission. To view a copy of this license, visit https://creativecommons.org/licenses/by-nc-nd/4.0

All versions of this work may contain content reproduced under license from third parties.

Permission to reproduce this third-party content must be obtained from these third-parties directly.

When citing this work, please include a reference to the DOI 10.1017/9781009383486

First published 2024

*A catalogue record for this publication is available from the British Library*

*A Cataloging-in-Publication data record for this book is available from the Library of Congress*

ISBN 978-1-009-38346-2 Paperback

Cambridge University Press & Assessment has no responsibility for the persistence or accuracy of URLs for external or third-party internet websites referred to in this publication and does not guarantee that any content on such websites is, or will remain, accurate or appropriate.

# Contents

| | | |
|---|---|---|
| *List of Figures* | | *page* vii |
| *List of Tables* | | viii |
| *Notes on Contributors* | | ix |
| *Acknowledgments* | | xviii |

1 Introduction: Studying the Role and Influence of International Environmental Bureaucracies    1
HELGE JÖRGENS, NINA KOLLECK, AND MAREIKE WELL

2 Means of Bureaucratic Influence: The Interplay between Formal Autonomy and Informal Styles in International Bureaucracies    27
MICHAEL W. BAUER, STEFFEN ECKHARD, JÖRN EGE, AND CHRISTOPH KNILL

3 The Evolution of International Environmental Bureaucracies: How the Climate Secretariat Is Loosening Its Straitjacket    57
THOMAS HICKMANN, OSCAR WIDERBERG, MARKUS LEDERER, AND PHILIPP PATTBERG

4 Environmental Treaty Secretariats as Attention-Seeking Bureaucracies: The Climate and Biodiversity Secretariats' Role in International Public Policymaking    73
MAREIKE WELL, HELGE JÖRGENS, BARBARA SAERBECK, AND NINA KOLLECK

5 Moving beyond Mandates: The Role of UNDP Administrators in Organizational Expansion    107
NINA HALL

| | | |
|---|---|---|
| 6 | Follow the Money: Secretariat Financing as a Window on the Principal–Agent Relationship<br>LYNN WAGNER AND PAMELA CHASEK | 131 |
| 7 | More Resources – More Influence of International Bureaucracies? The Case of the UNFCCC Secretariat's Clean Development Mechanism Regulation<br>KATHARINA MICHAELOWA AND AXEL MICHAELOWA | 153 |
| 8 | The Marrakech Partnership for Global Climate Action: Democratic Legitimacy, Orchestration, and the Role of International Secretariats<br>KARIN BÄCKSTRAND AND JONATHAN W. KUYPER | 180 |
| 9 | The Administrative Embeddedness of International Environmental Secretariats: Toward a Global Administrative Space?<br>BARBARA SAERBECK, HELGE JÖRGENS, ALEXANDRA GORITZ, JOHANNES SCHUSTER, MAREIKE WELL, AND NINA KOLLECK | 201 |
| 10 | Reflections on the Role of International Public Administrations in the Anthropocene<br>FRANK BIERMANN | 228 |
| *Index* | | 241 |

# Figures

| | | |
|---|---|---|
| 2.1 | Ideal-typical configurations of formal and informal potentials of bureaucratic policy influence | *page* 41 |
| 2.2 | Empirical configurations of formal and informal potentials of bureaucratic policy influence | 44 |
| 6.1 | Rio Conventions' core budgets plus trust funds | 141 |
| 6.2 | Contributors to the IPCC Trust Fund: 1989–2016 | 144 |
| 6.3 | IPCC Trust Fund contributions: 1989–2016 | 145 |
| 6.4 | Contributors to the IPBES Trust Fund: 2012–2017 | 146 |
| 6.5 | IPBES Trust Fund contributions: 2012–2017 | 146 |
| 7.1 | Development of financial and human resources of the UNFCCC Secretariat | 163 |
| 7.2 | CDM revenues for the secretariat and the development of the accumulated surplus over time (million USD) | 164 |
| 7.3 | CDM versus total secretariat staff | 165 |
| 7.4 | Changes in procedures for approval of baseline and monitoring methodologies over time | 168 |
| 8.1 | The O–I–T relationship | 183 |
| 9.1 | The combined CBD and UNFCCC network by actor groups | 211 |
| 9.2 | The combined CBD and UNFCCC network by UN conventions | 212 |
| 9.3 | Network of environmental bureaucracies | 216 |
| 9.4 | Network of environmental bureaucracies and their relations with state and nonstate actors by actor group | 219 |
| 9.5 | Network of environmental bureaucracies and their relations with state and nonstate actors by UN convention | 220 |

# Tables

| | | |
|---|---|---:|
| 2.1 | Measurement of bureaucratic autonomy | *page* 33 |
| 2.2 | Measurement of administrative styles | 38 |
| 2.A1 | Interview list for the measurement of administrative styles as presented in Figure 2.2 | 49 |
| 7.1 | Standardized baselines developed with Secretariat support | 170 |
| 7.A1 | Overview of the UNFCCC Secretariat's development as well as of its CDM- and JI-related activities | 173 |
| 9.1 | Top thirty organizations with the highest centrality values in the combined CBD and UNFCCC network | 213 |
| 9.2 | The thirty environmental administrative actors with the highest centrality values in the global environmental administrative space | 217 |
| 9.3 | The thirty organizations with the highest centrality values in the global environmental administrative space | 221 |

# Notes on Contributors

**Karin Bäckstrand** is a Professor in Environmental Social Science at the Department of Political Science at Stockholm University, a senior researcher at the Institute for Future Studies, and a member of the Swedish Climate Policy Council. Her research revolves around the democratic legitimacy of global environmental politics, non-state actors in climate change governance, and the role of public–private governance in sustainability governance. She previously held positions as Professor in Political Science, Lund University, visiting scholar at the University of Oxford, and Wallenberg fellow at the Massachusetts Institute of Technology. She leads a new four-year research program funded by the Swedish research council on sustainable development and the role of multistakeholder partnerships in the 2030 Agenda and the Sustainable Development Goals. Her work has been published in journals such as *Global Environmental Politics*, the *European Journal of International Relations*, *Global Environmental Change*, *Environmental Politics*, and the *Journal of European Public Policy*. She coedited *Rethinking the Green State: Environmental Governance towards Climate and Sustainability Transition* (with Annica Kronsell, Routledge, 2015) and *Research Handbook on Climate Governance* (with Eva Lövbrand, Edward Elgar, 2015). Her recent publications include *Legitimacy in Global Governance: Sources, Processes, and Consequences* (coedited with Jonas Tallberg and Jan Aart Scholte, Oxford University Press, 2018), and *Governing the Climate-Energy Nexus: Institutional Complexity and Its Challenges to Effectiveness and Legitimacy* (coedited with Fari Zelli et al., Cambridge University Press, 2020).

**Michael W. Bauer** holds the Chair of Public Administration in the School of Transnational Governance of the European University Institute, San Domenico di Fiesole, Florence. He also served as the Chair for Comparative Public Administration and Policy Analysis at the German University of Administrative Sciences Speyer (2012–2020), at the Humboldt University of Berlin (2009–2012), and at the University of Konstanz, Germany (2004–2009). He received his Ph.D. from the

European University Institute in 2000. His research focuses on European and international bureaucracy, populism and democratic administration, multilevel governance, and EU institutions. He publishes widely in international journals, among them the *Public Administration Review*, the *Journal of Public Administration Research and Theory*, *Public Administration*, the *Journal of European Public Policy*, the *Journal of Common Market Studies*, the *International Studies Review*, the *Journal of Comparative Policy Analysis: Research and Practice*, the *European Journal of Political Research*, and *Governance*.

**Frank Biermann** is a Professor of Global Sustainability Governance with the Copernicus Institute of Sustainable Development, Utrecht University, the Netherlands. He is the Director of the GlobalGoals research program on the steering effects of the Sustainable Development Goals, supported by a European Research Council Advanced Grant; the founder and first Chair of the Earth System Governance Project, a global transdisciplinary research network; editor-in-chief of the journal *Earth System Governance*; and coeditor of three book series with Cambridge University Press and MIT Press. His recent books – all with Cambridge University Press – are *Anthropocene Encounters: New Directions in Green Political Thinking* (coedited, 2019); *Architectures of Earth System Governance: Institutional Complexity and Structural Transformation* (coedited, 2020); and *The Political Impact of the Sustainable Development Goals: Transforming Governance through Global Goals?* (coedited, 2022). In 2021, the Environmental Studies Section of the International Studies Association presented him with its Distinguished Scholar Award.

**Pamela Chasek** is Professor of Political Science at Manhattan College in New York. She is also the co-founder and Executive Editor of the *Earth Negotiations Bulletin*, a reporting service on United Nations environment and development negotiations. She has written about and followed United Nations sustainable development negotiations for more than thirty years. Her research focuses on multilateral environmental negotiations and international environmental governance as a means for countries and nonstate actors to address and resolve environmental problems at the international level. She is the author and editor of numerous articles and books, including *The Roads from Rio: Lessons Learned from Twenty Years of Multilateral Environmental Negotiations* (Routledge, 2012), *Transforming Multilateral Diplomacy: The Inside Story of the Sustainable Development Goals* (Routledge, 2018), and *Global Environmental Politics, 8th edition* (Routledge, 2021).

**Steffen Eckhard** is Professor of Public Administration and Public Policy at Zeppelin University Friedrichshafen, Germany. He is also a fellow at the Cluster of Excellence "Inequality" at the University of Konstanz and a non-resident fellow at the Global Public Policy Institute in Berlin. Focusing on the management of public organizations, particularly in the international realm, Steffen contributes to advancing theories of public organizations and their multifaceted interactions with society and politics. His findings have been featured in the leading journals and book publishers of public administration and political science, including his most recent book *The Politics of Evaluation in International Organizations* (Oxford University Press, 2023).

**Jörn Ege** is Lecturer of Local and Regional Governance at the Zurich University of Applied Sciences in Winterthur, Switzerland. He was a Principal Investigator for the project "The Consequences of Bureaucratic Autonomy for International Administrative Influence" based at the German University of Administrative Sciences Speyer (2017–2021). His research interest is in the field of comparative public administration and public policy. He is particularly interested in social and health policy and its implementation at different territorial levels.

**Alexandra Goritz** is a Policy Advisor on climate foreign policy at Germanwatch and a doctoral student at the Free University Berlin. She holds an M.Sc. in Environment and Development from the London School of Economics and Political Science and a B.A. in Politics and Public Administration from the University of Konstanz. Her research is focused on the role of international organizations and their bureaucracies during global environmental negotiations. More specifically, she uses approaches involving social network analyses based on Twitter data to assess the authority of state and nonstate actors. She has published in *Climatic Change*, *Global Environmental Politics*, the *International Review of Administrative Sciences*, and the *Journal of Comparative Policy Analysis: Research and Practice*.

**Nina Hall** is an Assistant Professor of International Relations at Johns Hopkins School of Advanced International Studies. Her research explores the role of transnational advocacy and international organizations in international relations. Her latest book *Transnational Advocacy in the Digital Era* (Oxford University Press, 2022) won the International Studies Association Best Book Award in International Communication in 2023, was shortlisted for the British International Studies Association Susan Strange Best Book Prize (2023), and received an honorable mention from the American Political Science Association Information Technology & Politics Best Book Award (2023). Her first book was *Displacement, Development and Climate Change: International Organizations Moving beyond Their Mandates?* (Routledge, 2016). She has published on climate change and international organizations in the *European Journal of International Relations*,

*Global Environmental Politics*, and *Global Governance*. She previously worked as a lecturer at the Hertie School of Governance in Berlin and has a D.Phil. in International Relations from the University of Oxford. She is on the steering committee of an independent New Zealand think tank, Te Kuaka. She regularly publishes commentary in international media, including the *Guardian*, *Die Zeit Online*, the *Washington Post*, *Project Syndicate*, and *The Conversation*.

**Thomas Hickmann** is an Associate Senior Lecturer in the Department of Political Science of Lund University in Sweden. Prior to this, he was a postdoctoral researcher with the Environmental Governance group at the Copernicus Institute of Sustainable Development of Utrecht University. His research deals with the global governance of global environmental changes and related transboundary sustainability challenges (such as biodiversity loss, climate change, and land degradation). He is co-convener of the taskforce on the Sustainable Development Goals of the Earth System Governance Project and served from 2015 to 2021 on the steering committee of the Environmental Politics and Global Change working group in the German Political Science Association. His most recent book, *The Political Impact of the Sustainable Development Goals: Transforming Governance through Global Goals?* (coedited with Frank Biermann and Carole-Anne Sénit, Cambridge University Press, 2022), was published open access.

**Helge Jörgens** is Associate Professor at the Department of Political Science and Public Policy at Iscte – University Institute of Lisbon and an integrated researcher at CIES-IUL, the Centre for Research and Studies in Sociology, Lisbon. He holds a Ph.D. from the Free University Berlin. He was Principal Investigator (together with Nina Kolleck) for the ENVIPA, CONNECT and TRANSPACE research projects on the role and influence of international environmental bureaucracies, supported by the German Research Foundation. His current research interests include environmental, energy, and climate policies, the role and influence of international public administrations, and the diffusion and transfer of public policies. His recent books are *A Guide to EU Renewable Energy Policy: Comparing Europeanization and Domestic Policy Change in EU Member States* (coedited, Edward Elgar, 2017), *Understanding Environmental Policy Convergence: The Power of Words, Rules and Money* (coedited, Cambridge University Press, 2014), and the *Routledge Handbook of Environmental Policy* (coedited, Routledge, 2023).

**Christoph Knill** is Chair of Political Science and Public Administration at the University of Munich. He studied Public Administration and Political Science at the University of Konstanz and obtained his Ph.D. from the University of Bielefeld in 1994. He was a research associate at the Max Planck Institute for the Study of

Societies in Cologne (1994–1995), the European University Institute in Florence (1995–1998), and the Max Planck Project Group for the Study of Common Goods in Bonn (1998–2000). Before joining the department, he was Professor of Political Science at the Universities of Jena (2001–2004) and Konstanz (2004–2014). His main research interests lie in the areas of comparative policy analysis and comparative public administration. In this context, his main focus is on policymaking in the European Union and the analysis of processes of international policy convergence and policy diffusion, as well as research on policy implementation. With regard to these topics, his major thematic interest is environmental, social, and morality policies.

**Nina Kolleck** is Full Professor at the University of Potsdam, Germany. Prior to her current role, she held professorships at Leipzig University, RWTH Aachen University, and the Free University Berlin. Additionally, she has served as a visiting Professor at institutions such as the University of California, Berkeley, the University of British Columbia in Vancouver, and the Hebrew University of Jerusalem. She is Principal Investigator (together with Helge Jörgens) of the ENVIPA, CONNECT and TRANSPACE research projects on the role and influence of international environmental bureaucracies, supported by the German Research Foundation. She received an ERC grant in 2022 and the Award for Research Cooperation and Highest Excellence in Science from the Max Planck Society (2017–2018). She is a member of various scientific advisory boards, advises different federal ministries, and acts as an anonymous reviewer for central funding organizations. Her current research interests include social network analysis, the transnationalization of nonstate actors, and climate education and policies, as well as the diffusion and transfer of innovations. Her recent research has been published in journals such as the *Journal of European Public Policy*, *Global Environmental Politics*, the *International Review of Administrative Sciences*, the *Journal of Comparative Policy Analysis: Research and Practice*, *Social Networks*, *Environmental Politics*, the *Journal of Education Policy*, *Global Governance*, *Teaching and Teacher Education*, the *Journal of Educational Administration*, *Educational Management Administration & Leadership*, *Environmental Education Research*, and the *European Educational Research Journal*.

**Jonathan W. Kuyper** is an Associate Professor of Political Science at the University of Oslo. Prior to this, he was an Assistant Professor at Queen's University Belfast and has held visiting positions at the University of Oxford's Blavatnik School, Princeton University, and the European University Institute. His current research focuses on the role of nonstate actors in world politics, the nature of deliberation in international negotiations, and the democratic legitimacy of international organizations. He is currently working on these topics in the issue areas of climate change governance, international trade, and investment. His articles have appeared in journals such as the *American Political Science Review*, *Global Environmental Politics*, the *European Journal of International Relations*, the *European Journal of Political Research*, and *Environmental Politics*. His most recent coauthored book – *Deliberative Global Governance* – was published with Cambridge University Press in 2019.

**Markus Lederer** is a Professor of Political Science with a focus on International Relations at Technical University Darmstadt, Germany. His research interests cover international relations, security politics, and global climate politics; he focuses on the politics of green transformations in the Global South, the institutional aspects of setting up market mechanisms, and the role of forest governance. He currently leads a research project, "The Political Institutionalization of Climate Change Mitigation in Emerging Economies," which is funded by the German Research Foundation. His most recent publications have appeared in *Regulation & Governance*, *Environmental Politics*, *Ecological Economics*, the *Cambridge Review of International Affairs*, the *Journal of Energy Markets*, *Geopolitics*, and the *Journal of Environment and Development*.

**Axel Michaelowa** is a Senior Researcher at the Institute of Political Science of the University of Zurich, Research Director at the research institute Perspectives Climate Research, and Senior Founding Partner of the climate policy consultancy Perspectives. He was the lead author of the chapter on mitigation policies in the fourth and fifth assessment reports of the Intergovernmental Panel on Climate Change. He has participated in the United Nations Framework Convention on Climate Change (UNFCCC) negotiations since 1995 and served on the Clean Development Mechanism (CDM) Registration and Issuance Team of the CDM Executive Board between 2006 and 2013. Axel has supported the Conference of the Parties presidencies of Qatar and Mexico and various country delegations in UNFCCC negotiations. He is a member of the Executive Committee of the Adaptation Benefits Mechanism and a member of the board of the Climate Cent Foundation between 2005 and 2009. He has also worked on ten approved baseline methodologies and three approved standardized baselines under the CDM and engaged with capacity-building in over forty developing countries, ranging from Algeria to Yemen. He has been involved in work on nationally determined contributions and nationally appropriate mitigation actions in Algeria, Bhutan, India, Morocco, Peru, Rwanda, Saudi Arabia, Tanzania, Trinidad and Tobago, Tunisia, Uzbekistan, and Vietnam.

**Katharina Michaelowa** is Professor of Political Economy and Development at the University of Zurich. She studied Economics at the University of Mannheim and at Delhi School of Economics. Before joining the University of Zurich, she held positions at the Organisation for Economic Co-operation and Development in Paris and at the Hamburg Institute of International Economics. Her research focuses on the effectiveness of development cooperation as well as on the effectiveness of developing countries' own policies. In this context, she also examines developing countries' climate policies and their positions in international negotiations, as well as the role of national and international administrations and governance structures.

**Philipp Pattberg** is a Professor of Transnational Environmental Governance and Policy Head of the Environmental Policy Analysis Department at the Institute for Environmental Studies, Vrije Universiteit Amsterdam. He is Director of the interfaculty Amsterdam Sustainability Institute, a platform for interdisciplinary and transformative research at the university. His research interests include governing oceans, climate, and biodiversity, and questions around institutional complexity and change. His recent books are the *Encyclopedia of Environmental Governance and Politics* (coedited, Edward Elgar, 2015), *Environmental Politics and Governance in the Anthropocene* (coedited, Routledge, 2016), and *The Anthropocene Debate and Political Science* (coedited, Routledge, 2018).

**Barbara Saerbeck** works for Agora Energiewende as Senior Associate Key Questions, focusing on national and European policy. Previously, she analyzed the German electricity, gas, and heat markets at the Federation of German Consumer Advocacy Organizations as a consultant. In the course of her work in the ENVIPA research project on international climate and biodiversity policy at the Free University Berlin she used social network analyses, among others, to examine the influence of state and nonstate actors on international negotiations. The ENVIPA project was part of the research unit International Public Administrations, funded by the German Research Foundation. She studied Political and Administrative Sciences in Konstanz, Prague, and Brussels as well as International Relations in Berlin and Potsdam. In her dissertation for the Environmental Policy Research Centre of the Free University of Berlin she investigated the influence of the European Environmental Agency on European decision-making processes. She is also a lecturer at the Berlin School of Economics and Law.

**Johannes Schuster** is a Research Associate at Leipzig University in Germany. He was also a member of the CONNECT project that aimed to systematically analyze the influence of international treaty secretariats in global networks. His research focuses on international organizations in global governance, policy networks, social network analysis, and the diffusion of social innovations. In his dissertation he examined the roles of intergovernmental and nonstate actors in the implementation of the United Nations Convention on the Rights of Persons with Disabilities and inclusive education. He studied Educational Science and Educational Research at the University of Jena and the Free University Berlin, obtaining his Ph.D. from the latter institution. His research has been published in journals such as the *Journal of Comparative Policy Analysis: Research and Practice*, the *Journal of Education Policy*, and *Teaching and Teacher Education*.

**Lynn Wagner** is Senior Director of the Tracking Progress program at the International Institute for Sustainable Development (IISD), where she oversees projects that seek to increase the accountability of decision-making for the Sustainable Development Goals and to support the implementation of the goals. Wagner began working with IISD in 1994 as a writer for the *Earth Negotiations Bulletin* and she continues to observe and analyze multilateral environmental negotiations for IISD's flagship publication. She teaches classes on bargaining and negotiation as an Adjunct Professor at the Johns Hopkins University's Paul H. Nitze School of Advanced International Studies. Her research and publications focus on the relationship between negotiation processes, outcomes, and justice. Her publications include "The Role of Issues in Negotiation: Framing, Linking, and Ordering" (coauthored, *Negotiation Journal*, 2021), "Justice Matters: Peace Negotiations, Stable Agreements, and Durable Peace" (coauthored, *Journal of Conflict Resolution*, 2017), and *The Roads from Rio: Lessons Learned from Twenty Years of Multilateral Environmental Negotiations* (coedited, Routledge, 2012).

**Mareike Well** is a Ph.D. candidate at the Free University Berlin. She works in the office of the Minister of State for Europe and Climate in the German Federal Foreign Office on international and European climate policy and climate diplomacy. Previously, she was a research fellow in the ENVIPA project and worked for the German Advisory Council on the Environment as well as for the German Federal Ministry for the Environment, Nature Conservation, and Nuclear Safety. Her research interests include multilateral negotiations on climate change, biodiversity, and educational issues, as well as the role of international public administrations in global environmental governance. Her recent publications have appeared in *Global Governance*, *Global Environmental Politics*, and *Climate Policy*.

**Oscar Widerberg** is Deputy Department Head and Associate Professor in Environmental Policy and Politics in the Environmental Policy Analysis section at the Institute for Environmental Studies at the Free University Amsterdam. His research focuses on the role of cities, regions, companies, and other nonstate and subnational actors in global governance. His work has been featured in academic articles, books, policy reports, and opinion pieces, including in *Nature Climate Change*, *Environmental Politics*, and *Millennium*. In 2019, he was the lead author of the United Nations Environment Programme gap report chapter and the stand-alone report on nonstate and subnational climate action. Prior to joining the Institute for Environmental Studies, he worked in consulting, advising international public authorities, primarily European institutions, on energy, environment, and climate policies.

# Acknowledgments

First of all, we thank the authors of this volume for their contributions and for their patience with the coming-into-being of this book. A warm thank you also goes to the anonymous reviewers from Cambridge University Press, who provided excellent feedback, input, and ideas, all of which increased the quality and accessibility of the book. Robert O. Keohane read and commented on an early version of this book. We are thankful for his constructive feedback. We thank Frank Biermann for including this book in his excellent Earth System Governance Series at Cambridge University Press. We furthermore wish to thank our research assistants, Berfin Yildirim and Christiane Emmerling, for their precise and steadfast help in formatting the manuscript and supporting our administration. We are grateful for the kind guidance received from Cambridge University Press staff, especially Emma Kiddle, Sarah Armstrong, and Matt Lloyd. Finally, we thank the German Research Foundation, to which we are greatly indebted for funding a large-scale research program on the study of international public administrations that has been carried out over the last ten years (www.ipa-research.com), including the ENVIPA research project (German Research Foundation funding codes: JO 1142/1-1 and KO 4997/1-1), the CONNECT research project (German Research Foundation funding codes: KO 4997/4-1 and JO 1142/2-1), and the TRANSPACE project on the emergence of transnational administrative spaces in environmental governance (German Research Foundation funding code: KO 4997/10-1). This book combined central findings from the scholars active in these projects with findings of some of the most renowned international scholars in the field of global environmental policy. We are grateful for this opportunity to jointly showcase the broad range of new conceptual, empirical, and methodological insights into the role of international public administrations in global environmental governance.

# 1

# Introduction

*Studying the Role and Influence of International Environmental Bureaucracies*

HELGE JÖRGENS, NINA KOLLECK, AND MAREIKE WELL

## 1.1 Introduction

In recent years, scholars of public administration and international relations have increasingly turned their attention to the role and impact of international public administrations (IPAs), that is, the bureaucratic bodies of international organizations (IOs) (Bauer 2006; Bauer, Knill, and Eckhard 2017a; Biermann and Siebenhüner 2009b; Knill and Bauer 2016; Lenz et al. 2015). Within this research strand, there has been particular focus on the secretariats of multilateral environmental conventions as potentially influential actors in world politics (Biermann and Siebenhüner 2009b; Jinnah 2014) and the degree to which these can act autonomously, that is, beyond the direct control of a treaty's member states (Bauer and Ege 2016; Eckhard and Ege 2016; Mathiason 2007). Moreover, scholars have started to explore the extent to which treaty secretariats are able to exert autonomous influence on the processes, outputs, and implementation of multilateral treaty negotiations as well as the causal mechanisms through which this influence is exercised (Biermann et al. 2009; Depledge 2007; Jinnah 2011; Knill and Bauer 2016: 950–956).

A milestone in this research was the study by Biermann and his colleagues (Biermann and Siebenhüner 2009b) that described international environmental bureaucracies as active and consequential "managers of global change." The study identified three mechanisms through which these bureaucracies were able to influence the formulation and implementation of international environmental policies – cognitive, normative, and executive influence – and argued that the degree of influence depended to an important extent on the leadership and staff of international bureaucracies and their attitudes, preferences, and strategies (Biermann and Siebenhüner 2009a). Although the study dove deep into the role and influence of international environmental secretariats, it left some questions unanswered and raised a multitude of new ones, thereby setting the stage for an important and fruitful research program that brought about important insights into the institutions, processes, and actor constellations of global environmental governance as a whole.

With this book, we would like to advance the debate on the influence of IPAs, answer some of the most important and still open questions, and outline how this lively field of research has evolved over a decade after the publication of the seminal work of Biermann and Siebenhüner (2009b). This book brings together contributions from many of the most renowned scientists in the field, presents new answers and research findings, and identifies current research gaps and perspectives for future research in an increasingly relevant field.

In this introduction, we not only review the scholarly literature that has followed the direction of Biermann and Siebenhüner (2009b) but also link it to some of the very early predecessors of the current IPA research agenda. Section 1.1 defines IPAs and distinguishes them from the wider IOs or treaty systems that they are an integral part of. Section 1.2 briefly addresses the question of whether and how IPAs should be expected to matter in global governance. Section 1.3 gives some examples where IPAs were found to have had an autonomous and discernible influence on international policy processes and outputs. Section 1.4 then asks for the determinants of IPA influence, gives an overview of the most relevant causal factors, and outlines how the chapters in this book contribute to the research on IPA influence.

## 1.2 From IOs to IPAs: Defining the Object of Analysis

In 1971, in a special issue of the journal *International Organization*, Robert O. Keohane and Joseph S. Nye diagnosed what they called a "Mount Everest syndrome" in the study of IOs. They argued that scholars were studying international organizations simply because "they are there," not because they actually mattered (Keohane and Nye 1971: v). This harsh criticism marked the beginning of a period of scholarly neglect of IOs as actors in their own right. IOs were mainly conceived of as abstract sets of rules designed by states to facilitate intergovernmental cooperation (Keohane 1984; Martin and Simmons 2012). Only in the late 1990s did researchers begin to rediscover earlier conceptualizations of IOs as agents in their own right and to systematically study their role in world politics and their influence on international policy outputs (Barnett and Finnemore 2004; Biermann and Siebenhüner 2009b; Hawkins and Jacoby 2006; Reinalda and Verbeek 1998). Rooted in theoretical frameworks such as principal–agent theory, sociological institutionalism, and other organizational theories, these studies have left the Mount Everest syndrome behind, allowing political science scholars to study IOs not merely because they are there but because there is strong theoretical and empirical evidence that they actually matter, not just as sets of rules but also as actors in their own right who are involved in processes of global policymaking.

Scholars utilizing a principal–agent perspective make the functionalist argument that nation-states (principals) delegate powers to IOs (agents) when they fail to coordinate directly (Abbott and Snidal 1998; Tallberg 2010). Governments expect IOs to carry out only those tasks that are deliberately delegated to them. However, owing to incomplete delegation contracts and information asymmetries, IOs may increase their organizational autonomy and begin to pursue agendas of their own (Bauer and Weinlich 2011: 254). International civil servants who successfully manage to influence the mandate and institutional design of newly established intergovernmental organizations represent just one of many examples of this extension of autonomy (Johnson and Urpelainen 2014). From a principal–agent perspective, the influence of IOs is thus a direct result of their autonomy from member states, and the degree of autonomy is a function of the latter's limited ability to control and sanction the former (Liese and Weinlich 2006: 504). Consequently, principal–agent theorists explain varying degrees of IO influence primarily through differences in principal preferences, constellations, and decision rules (Da Conceição, 2010; Hawkins et al. 2006a), paying less attention to factors inherent to IOs. Principal–agent approaches are thus most effective in explaining differences in IO influence when external factors differ between cases. Where principal preferences and constellations are constant and varying degrees of IO influence persist, principal–agent theory has less insight to offer (Hawkins and Jacoby 2006).

Sociological institutionalism fills this gap by focusing on factors inherent to IOs as sources of administrative influence. Sociological institutionalism, in particular the "bureaucratic authority" variant employed by Barnett and Finnemore (2004), focuses on the normative and cultural roots of the influence of IOs (Fleischer and Reiners 2021). From this perspective, IOs become influential owing to their expertise, institutional memory, moral standing, and – based on these factors – their privileged position in social networks (Wit et al. 2020). IOs know more about technical and legal issues than their political masters (Derlien, Böhme, and Heindl 2011: 91) and have superior "informal knowledge about the history and evolution of institutional processes" (Jinnah 2010: 62; see also Biermann et al. 2009; Dijkstra 2010; Jinnah 2014). This bureaucratic authority (Hickmann 2019) of IOs forms the basis of their influence on processes of international rulemaking. Their political standing is further enhanced by their claim to defend the international common good based on scientific expertise (Busch and Liese 2017; Busch et al. 2021; Herold et al. 2021) rather than pursue vested interests. As a result, IOs and their bureaucracies try to uphold a reputation for neutrality by avoiding any impression that they are pursuing their own agenda (Barnett and Finnemore 2004: 21).

While the reconceptualization of IOs as political actors in their own right builds on the implicit distinction between IOs on the one hand and their bureaucratic bodies or secretariats on the other, this distinction is not always made explicit and is still far from omnipresent in the field of international relations. As Weinlich (2014: 39) puts it: "Most of the recent literature does not bother to make a distinction between international organisations and their bureaucracies. Often, scholars who are referring to international organisations as actors ... are actually, albeit rarely explicitly, referring to the respective bureaucracy." Similarly, Eckhard and Ege (2016: 967), in their systematic review on how international bureaucracies influence the policies of IOs, conclude that only few studies "explicitly focus on the influence of IPAs as a dependent variable." Findings regarding the "bureaucratic footprint" in the policies of IOs are "a side-product rather than the actual objective of most studies." In order to more systematically study to what extent and through which causal mechanisms international bureaucracies can shape international policy outputs, IPAs must be treated as actors that are analytically distinct from the wider international organization or treaty system that they are a part of.

This distinction between IOs and their bureaucratic bodies has been most clearly made in the field of organizational studies. Organizational perspectives on IOs explicitly attribute explanatory power to the organizational features of the bureaucratic bodies of IOs: organizational design, secretariat leadership, and shared preferences among international civil servants (Jönsson 1986; Ness and Brechin 1988). Organizational design comprises the "formalized internal rules and procedures that assign tasks and positions in the hierarchy." When these are poorly specified, "conflicts, redundancies, inefficiencies, [and] delays" might ensue (Biermann et al. 2009: 55). Whether organizational structure actually influences international policy outputs depends to a great extent on the leadership provided by the IPA's top management, whose convictions regarding the role bureaucracy should play in international policymaking can vary considerably (Depledge 2007: 63; Siebenhüner 2009: 268; Siotis 1965). Strong leadership by executive secretaries "that is charismatic, visionary, and popular, as well as flexible and reflexive" is assumed to enhance a bureaucracy's effectiveness by increasing internal and external acceptance of and trust in top management and its abilities (Biermann et al. 2009: 58). Finally, the governance preferences of the international civil service – for example, whether civil servants value active political engagement as opposed to passive neutrality – may also account for varying levels of IPA influence (Bauer 2006: 44; Busch 2009a: 258).

Since the end of the first decade of the twenty-first century, a rapid convergence of these formerly distinct research agendas on international bureaucracies has been observed (Bauer et al. 2017; Biermann and Siebenhüner 2009b; Busch 2014; Dijkstra 2017; Ellis 2010; Fleischer and Reiners 2021; Trondal 2017; Wit et al. 2020). While in 2009, Biermann and Siebenhüner (2009c: 1) had still found it "remarkable" that in the academic field of international relations "the scholarly study of the influence of international bureaucracies has been a rather peripheral research object for most of the post-1945 period," Trondal (2017: 35), less than a decade later, characterized IPAs as "a distinct and increasingly central feature of both global governance studies and public administration scholarship" (see also Martin and Simmons 2012: 329). These "*separate* international administrations that are able to act relatively independently from domestic governments" (Trondal 2017: 37, emphasis in original) are now seen to constitute a central and analytically distinct component of any attempt to build a common political order at the international level.

### 1.3 What Are International Bureaucracies?

Biermann and Siebenhüner (2009c: 6) define international bureaucracies "as agencies that have been set up by governments or other public actors with some degree of permanence and coherence and beyond formal direct control of single national governments (notwithstanding control by multilateral mechanisms through the collective of governments) and that act in the international arena to pursue a policy." While the authors follow earlier characterizations of IPAs, such as that of Siotis (1965: 178), who defined IPAs as "international bodies which have a distinct existence within a given system of multilateral diplomacy and which exercise administrative and/or executive functions, implicitly recognized or explicitly entrusted to them by the actors of the international system," they place greater emphasis on the autonomy and actorness of these organizations. This emphasis on autonomy is also taken up by Bauer et al. (2017: 2), who describe IPAs "as bodies with a certain degree of autonomy, staffed by professional and appointed civil servants who are responsible for specific tasks and who work together following the rules and norms" of a given international organization.

Most IPAs are "issue-specific" bureaucracies (Bauer 2006: 28). Except for the secretariats of universal IOs, such as the United Nations Secretariat, their functions are usually closely related to a policy domain or to the topic of a multilateral treaty. Within these policy domains, IPAs engage in activities "such as conducting studies, preparing draft decisions…, assisting states parties, and receiving reports on the implementation of commitments" (Churchill and Ulfstein 2000: 627). Their tasks "typically range from generation and processing of data, information and knowledge over providing administrative, technical, legal and advisory support in intergovernmental negotiation processes to ensuring and monitoring compliance with multilateral decisions" (Busch 2014: 46–47). In 1994, Sandford (1994: 19) argued that international secretariats invariably act in a servant-like fashion: "Underlying all secretariat activities is the notion of service. Secretariats exist to service the treaty parties." More recent research by Knill et al. (2018), however, shows that the servant-like IPA is just one among several possibilities. Distinguishing different administrative styles of IPAs, the authors show that the servant style is no longer the default behavior of international bureaucracies but that IPAs may just as well adopt entrepreneurial or even advocacy-oriented administrative styles. While these styles may vary between different IPAs, they may also vary across issue areas or phases of the policy cycle (Bayerlein, Knill, and Steinebach 2020; Knill et al. 2018; see also Well et al. 2020). This diversity of administrative styles indicates that – despite not having any formal decision-making powers – IPAs often attempt to move beyond the role of passive servants and to influence the processes and outputs of their respective IOs or treaty systems. Against this backdrop, Trondal (2017: 36) sums up: "It has been shown that the task of IPAs has become increasingly that of active and independent policy-making institutions and less that of passive technical supply instruments for IGO plenary assemblies."

Consequently, much of the more recent scholarly literature on IPAs has focused on whether and through which causal mechanisms international bureaucracies can become influential actors in international politics.

### 1.4 Examples of IPA Influence

While there is little doubt that IPAs may have an autonomous influence on international policy processes and outputs, concrete examples of IPA influence are still relatively scarce. A main reason is the methodological challenges of observing the often-hidden activity of IPAs. In addition, it is often methodologically difficult to link the actions of IPAs to observed changes in the processes or outcomes of multilateral negotiations. The fact that IPAs either do not reveal their political preferences or pass them off as preferences of other actors makes it even more difficult to clearly identify IPA action (or the preferences of IPAs) as the cause of observed policy changes.

A prominent example of IPA influence was the role UN Secretary-General Kofi Annan and his bureaucracy played in developing the principle of a "Responsibility to Protect." Characterizing the UN Secretary-General as an international norm entrepreneur, Johnstone (2007: 124) argues that this strategy "is likely to be most effective when he uses the United Nations to crystallize emerging understandings among states and non-state actors, rather than striking out in entirely new normative directions." In an earlier study, Bhattacharya (1976) found that the Secretariat of the United Nations Conference on Trade and Development (UNCTAD) significantly contributed to the agreement on the Generalized System of Preferences that was reached in 1970. The factors that enabled the secretariat to become influential were secretariat ideology, charismatic leadership by UNCTAD's Secretary-General Raul Prebisch, and coalition-building activities by secretariat staff.

IPA influence may also be relatively high in newly emerging policy domains. Levinson and Marzouki (2016: 70), in their study of the role of IOs in the field of global internet governance, observe that the secretariats of the Organisation for Economic Co-operation and Development (OECD), the Council of Europe, and the United Nations Educational, Scientific and Cultural Organization (UNESCO) played "a role in crafting ideas, first to be adopted by the member states and then disseminated externally, often with 'allies' or 'partners.'" They find that the UNESCO Secretariat developed the idea of "internet universality," the Secretariat of the Council of Europe ensured a stronger emphasis on human rights and stakeholder participation, and the OECD Secretariat was responsible for a stronger shift toward data protection in global internet governance.

In the environmental field, a first research strand focused predominantly on individual bureaucracies such as the OECD environmental directorate (Busch 2009b), the biodiversity secretariat (Siebenhüner 2007, 2009), and the World Bank Environment Department (Gutner 2005; Nielson and Tierney 2003). For example, Bauer (2009: 300) shows that "the desertification secretariat was pivotal in the establishment" of a permanent subsidiary body for implementation, the Committee for the Review of the Implementation of the Convention. This subsidiary body was established against the preferences of most donor countries. The example shows how treaty secretariats can actively shape the international institutions they are supposed to serve rather passively. Siebenhüner (2009: 272) finds that the biodiversity secretariat has traditionally been "entrusted with the drafting of decisions of the conference of the parties." While in highly contested issue areas these drafts provided by the secretariat were usually amended or rewritten by the negotiation parties, secretariat proposals on more technical issues often passed with only minor amendments.

Building on this research, a second wave of case studies linked the study of environmental bureaucracies to current research topics from a range of political science subdisciplines such as international relations and international public administration. Examples are studies on how treaty secretariats deal with the institutional fragmentation of global governance (Jinnah 2014), questions of delegation and agency in global environmental politics (Wagner and Mwangi 2010), or studies on the interplay of public and private governance at different levels of government (Chan et al. 2015; Dingwerth and Jörgens 2015; Newell, Pattberg, and Schroeder 2012). For example, focusing on institutional fragmentation, Jinnah (2012: 113) finds that "nearly all tools" used by the conferences of the parties of the Convention on Biological Diversity (CBD) "to mandate overlap management activities can be traced back to one document produced by the Secretariat in 1995." This example shows that IPA input may create path dependencies that perpetuate individual instances of IPA influence over longer periods.

Recently, innovative methodological approaches, combining quantitative social network analysis (SNA) with qualitative case studies, have been developed to overcome the methodological challenge of identifying the policy preferences of international secretariats. By focusing on issue-specific information flows between international bureaucracies and other actors in the global climate and biodiversity policy networks, these studies offer the potential to look behind the scenes of multilateral environmental negotiations and to trace the policy outputs of IOs or multilateral treaty systems back to IPA action (Goritz, Jörgens, and Kolleck 2021, 2022; Goritz et al. 2020; Jörgens, Kolleck, and Saerbeck 2016; Kolleck et al. 2017; Mederake et al. 2021).

Albeit incomplete, this selection of examples illustrates some of the many potential sources of IPA influence. The next section provides a systematic review of the literature on factors that potentially affect the ways and extent to which international bureaucracies can influence international policy outputs.

## 1.5 Determinants of IPA Influence

Already in 1974, Keohane and Nye (1974: 52) argued that "[m]ost intergovernmental organizations have secretariats, and like all bureaucracies they have their own interests and goals that are defined through an interplay of staff and clientele. International secretariats can be viewed both as catalysts and as potential members of coalitions; their distinctive resources tend to be information and an aura of international legitimacy." More recently, and based on a set of case studies, Bauer, Knill, and Eckhard (2017b: 182–189) distinguish five sources of IPA influence: First, and contrary to an instrumental view that conceives of IPAs as mere instrumental arrangements created to support intergovernmental cooperation, they argue that IPAs are inherently autonomous and even more so than their national counterparts (see also Bauer and Ege 2017). Second, they find that IPAs are entrepreneurial, meaning that they use their autonomy to advocate their own policy ideas and preferences (see also Jörgens et al. 2017; Knill et al. 2017). Third, expertise and information are more important tools for IPAs than rules and formal powers. While the formal mandates and legal competencies of IPAs are rather limited when compared with those of national bureaucracies, their strategic use of expertise, ideas, and procedural knowledge combined with their often central position in issue-specific information flows (nodality) forms the basis of their impact on global policy outputs (see also Busch and Liese 2017). Fourth, IPAs are able to overcome budgetary restrictions by generating new sources of financing. Although IPAs are much more vulnerable to budgetary instability than national bureaucracies, they find ways of mobilizing "budgetary means from alternative sources in order to reduce their dependence on member state contributions" (Bauer, Knill, and Eckhard 2017b: 187; see also Patz and Goetz 2017). Finally, the authors find that IPAs actively shape their organizational environment. They do so by setting up and forming structures of multilevel administration and by creating informal alliances with nonstate actors at all levels of government. IPAs then typically occupy a central position in "their" domain-specific organizational environment, especially within domain-specific information flows (see also Benz, Corcaci, and Doser 2017; Jörgens, Kolleck, and Saerbeck 2016). With an explicit focus on international environmental bureaucracies, Wit et al. (2020) identify three general sources of IPA influence: their degree of organizational autonomy, their ability to deliver specific governance functions, and the way in which the complex multilevel and multiactor structure of the international system enables IPAs to become active participants in processes of global governance. In the following, we will zoom in on some of these potential determinants of IPA influence.

### *Autonomy from Their Principals*

Verhoest et al. (2010: 18–19) define autonomy as "the extent to which an agency can decide itself about matters that it considers important." With regard to IOs and IPAs, Hawkins et al. (2006b: 8) define autonomy as "the range of potential independent action available to an agent after the principal has established mechanisms of control.... That is, autonomy is the range of maneuver available to agents after the principal has selected screening, monitoring, and sanctioning mechanisms intended to constrain their behavior." The autonomy of IPAs is mainly defined by the amount of discretion that the member states of an international organization or treaty system decide to grant their bureaucracy. Bauer and Ege (2016) refer to this as an IPA's "formal autonomy" (see also Chapter 2).

But the initial delegation of a certain degree of autonomy to an IPA through formal mandates is not the only factor that determines the bureaucracy's range of maneuver. Various other factors have been found to affect an IPA's autonomy. The first one is the fact that IOs and their IPAs are formal organizations whose "organizational development" (Schmitter 1971) cannot fully be controlled from the outside. Schmitter (1971: 918), building on Keohane's (1969) notion of institutionalization, describes organizational development of IOs as

> a process whereby an initially dependent system, created by a set of actors representing different and relatively independent nation-states, acquires the capabilities of a self-maintaining and self-steering system. Any system with such emergent properties remains, of course, related to and interdependent with its environment, but it becomes increasingly flexible, i.e., it is able to survive changes in that environment, and autonomous, i.e., "[its] course cannot be predicted from knowing only [its] environment."

Against this backdrop, we distinguish between the delegated or *formal* autonomy of an IPA and the autonomy resulting from its internal organizational strategies and development, which can be referred to as its *organizational* autonomy.

Several factors can affect an IPA's organizational autonomy. The first one has to do with the structure of the principal–agent relationship that is typical for international bureaucracies. In IOs and treaty systems, a bureaucracy's principal is often less homogeneous than at the national level. Vaubel (2006), for example, argues that international bureaucracies tend to be more autonomous from their principals than their national counterparts because the chains of delegation are longer and more complex. As Dehousse (2008) points out, international bureaucracies are normally controlled by multiple principals. Distinguishing "multiple" from "collective" principals, Dijkstra (2017: 603) describes the consequences for IPA autonomy as follows (see also Nielson and Tierney 2003): "We speak of a collective principal when the member states collectively interact with an agent. In the case of multiple principals, member states also unilaterally interact with the agent." If an international organization or treaty system is characterized by multiple principals, there is a potential chance for secretariats to team up with selected states with whom they share some interests against the interests of other states. Multiple principals may thus strengthen a secretariat's organizational autonomy and constitute a potential precondition for secretariat influence beyond their formal mandate. In contrast, as Jönsson (1986: 44) points out, "hegemonic and polar issue structures, where issue-specific capabilities are concentrated in one or a few states, can be expected to allow less room for maneuver by IOs than fragmented structures."

The increased organizational autonomy of international secretariats does not just become visible in their influence on multilateral policy outputs. International bureaucracies are also important actors in the process of creating new IOs or redefining, and often expanding, the mandates of existing ones (Johnson 2013, 2014; Johnson and Urpelainen 2014). For example, Johnson (2014: 6) shows that "[i]nternational bureaucrats working in pre-existing IGOs can – and do – advocate the creation of new institutions, participate in the institutional design process, and dampen the mechanisms by which states endeavor to control new institutions." The fact that the majority of IOs created in the past five decades are so-called emanations, that is, IOs that were created not by states but by other IOs (Pevehouse, Nordstrom, and Warnke 2005; Shanks, Jacobson, and Kaplan 1996), opens a potential new and institutional sphere of influence for international bureaucracies. Similar dynamics might also occur when an IPA attempts to redefine or expand its own mandate (see, e.g., Barnett and Coleman 2005). Against this backdrop, Chapter 5 by Nina Hall analyzes how and to what extent the United Nations Development Program (UNDP) was successful in integrating climate adaptation into its mandate. Hall argues that UNDP administrators, rather than states, played a critical role in mandate expansion by deciding "*whether* and *how* to expand into a new issue area" and then lobbying states to endorse this expansion. The chapter contributes to an emerging literature on how the leaderships of IPAs navigate financial, ideational and normative opportunities to expand their bureaucracies' mandates. Chapter 7 by Katharina Michaelowa and Axel Michaelowa argues that IPAs may also profit from new sources of revenue within their treaty systems. The authors show that the increased revenue from the Clean Development Mechanism (CDM) within the United Nations Framework Convention on Climate Change (UNFCCC) both directly and indirectly strengthened the role of the climate secretariat. Conversely, when this revenue decreased, the UNFCCC Secretariat lost part of its autonomous regulatory influence on the CDM and "tried to reorient CDM resources for support of the Paris Agreement negotiations and implementation of national mitigation action."

Another factor that may affect an IPA's organizational autonomy is salience or visibility of its actions. As Finkelstein (1974: 501) observed already in 1974, "[i]nstitutional autonomy correlates with lack of salience to the powerful members." Consequently, many studies find that IPAs attempt to maintain an image of neutrality, deliberately hiding their own policy preferences behind those of their IO's or treaty system's member states or other actors. If IPAs attempt to influence multilateral negotiations, they often do so in an "invisible" "or behind the scenes" way (Bauer 2006: 32; see also Well et al. 2020). Mathiason (2007), for example, refers to the political influence of international secretariats as "invisible governance." Jinnah (2014) writes that "[f]rom the outside of an organization, office secretaries are nearly invisible." With regard to the World Trade Organization, Bohne (2010: 116) finds that "[i]nfluences of the Secretariat and of chairpersons on the substance of negotiations are hidden, informal, and highly contingent upon times and personalities." In addition, Beach (2008: 220) cites an official of the General Secretariat of the Council of the European Union saying that "[l]e Secrétariat du Conseil n'existe pas."

However, maintaining a low-key profile is not the only way in which IPAs can increase their organizational autonomy. IPA scholars increasingly observe that international secretariats step out from behind the scenes and put themselves in the spotlight of multilateral negotiations, side by side with their principals and a range of nonstate and substate actors. A case in point is the secretariat of the UNFCCC. In 2009, Busch found that the climate secretariat was caught in a "straitjacket" of "formal and informal rules" imposed by the UNFCCC member states that "ruled out any proactive role or autonomous initiatives" and led to an "organizational culture that bars staff ... from exercising any leadership vis-à-vis parties and from assuming a more independent role" (Busch 2009a: 261). Today, this characterization no longer seems accurate as several scholars consider that the climate secretariat is "loosening its straitjacket" (see Chapters 3 and 7). In reaction to the failure of a globally binding post-Kyoto agreement on climate change at the UN Conference of the Parties (COP15) in 2009 in Copenhagen (Dimitrov 2010) and confronted with long-lasting stalemate among the formal negotiating parties, the UNFCCC Secretariat no longer acts as a passive servant to the negotiating parties. Instead, it has increasingly turned its attention to other nonparty actors at different levels of government in order to gain leverage on the substance and processes of global climate governance.

This changing role of international environmental treaty secretariats is reflected in new concepts of IPAs as orchestrators (Abbott and Snidal 2010; Abbott et al. 2015; also see Chapters 3 and 8) or as attention-seeking bureaucracies (see Chapter 4). For example, Bäckstrand and Kuyper (2017: 765) argue that "a crucial outcome of the Paris Agreement is that the UNFCCC has been consolidated as the central orchestrator of non-state actors and transnational initiatives in global climate governance." Jörgens et al. (2017) suggest that IPAs may attempt to strengthen their autonomy by actively attracting the attention of policymakers in order to feed their own policy-relevant knowledge and preferred policy recommendations into multilateral negotiations. Both concepts argue that the complex and dynamic institutional structure of multilateral agreements provides the organizations acting inside them with multiple options for strategic positioning (on the opportunity structure provided by environmental treaty systems, see Gehring 2012). In these cases, the underlying logic of action of international bureaucracies shifts from "shirking" to "attention-seeking." Interestingly, the possibility that "international secretariats or components of secretariats" could "form explicit or implicit coalitions with subunits of governments as well as with nongovernmental organizations having similar interests" had already been suggested by Keohane and Nye (1974: 52).

To the extent that international environmental bureaucracies develop their own policy preferences and are able to feed them into international and national decision-making processes, this influence may raise problems of democratic legitimacy. In Chapter 8, Karin Bäckstrand and Jonathan W. Kuyper analyze the normative problems associated with the practice of orchestration by international secretariats. The authors argue that "orchestration engenders a democratic duty" on the orchestrator "to ensure that their own actions, and those of intermediaries, are democratically legitimated by those affected, including both targets and additional actors implicated in the orchestration relationship." They illustrate their argument with an empirical case study of the UNFCCC Secretariat's orchestration efforts in the context of the Marrakech Partnership for Global Climate Action.

### *New Conceptualizations of IPAs' Organizational Autonomy*

Which concepts and theories can best describe the changed role and strategies of international bureaucracies in an international environmental and climate policy arena characterized by institutional fragmentation (Keohane and Victor 2011;

Zelli and Asselt 2013) and a diversification of actors (Hale and Roger 2014)? What are the implications of this reconceptualization for the future analysis of secretariat behavior? Against this backdrop, Michael W. Bauer et al. in Chapter 2 develop a model to explain why and how IPAs become influential actors in world politics. The authors base their model on the concepts of structural autonomy and administrative styles and lay out a strategy for their measurement. Based on these two measures, which represent the formal (structural autonomy) and informal (administrative styles) sources of IPA influence, they compare the empirical pattern of autonomy and style in a sample of eight administrations. The chapter concludes by putting forward propositions about the potential consequences of typical combinations of autonomy and style for international bureaucratic influence.

Chapter 3 by Thomas Hickmann et al. studies the UNFCCC Secretariat's proactive role in bringing nonstate actors that are supportive of the secretariat's policy preferences into the UNFCCC negotiations. It does so, for example, through secretariat-led initiatives such as the Lima–Paris Action Agenda or the Non-state Actor Zone for Climate Action. While Hickmann et al. base their case study on the concept of IOs as orchestrators, Chapter 4 by Mareike Well et al. proposes to conceive of international secretariats as attention-seeking bureaucracies. Well et al. argue that in order to become influential, international bureaucracies need to not only possess policy-relevant expert knowledge but also exploit the complex structures and actor constellations of multilateral treaty systems in ways to make negotiators take notice and adopt some of the bureaucracy's policy positions. The authors argue that in order to influence the outcomes of multilateral negotiations, international secretariats need to actively and strategically seek to attract the attention of the negotiating parties to their preferred problem definitions and policy prescriptions. This argument is illustrated with two case studies on the strategic behavior of the secretariats of the CBD and the UNFCCC.

### *Centrality of IPAs within Multilateral Negotiation Systems*

Besides its formal and organizational autonomy, an IPA's influence on negotiation processes and outputs is also characterized by its centrality in issue-specific policy networks. As Sandford (1994: 17) observed, "[s]ecretariats are the organizational glue that holds the actors and parts of a treaty system together." Similarly, Jinnah (2012: 109) characterizes secretariats as "the operational hubs of [their] regimes." This centrality allows IPAs to interact with a wide range of actors and potentially occupy a brokerage position between actors who do not interact directly with each other. Jönsson (1986: 45) refers to this as a "linking-pin position" and highlights that "[i]n order to assume an effective linking-pin position, an organization needs to have a location in the issue-specific network which allows it to reach, and to be reached by, other important organizational actors. Multiplexity of direct and indirect links with these actors can be expected to enhance the leverage of the prospective linking-pin organization." In a similar vein, Fernandez and Gould (1994: 1460) argue that "organizational actors linking otherwise unconnected pairs of actors play a critical role in policy domains because they permit information to flow easily among a large and diverse set of actors, which in turn allows actors to coordinate their efforts to formulate and influence policies." In their study of influence in the US health policy domain, the authors find "that occupancy of brokerage positions in the network of communication among organizational actors is positively related to influence" (Fernandez and Gould 1994: 1456).

In the environmental field, several studies have shown that treaty secretariats such as the climate or biodiversity secretariats occupy very central positions both in issue-specific communication flows and in issue-specific cooperation networks (Goritz, Kolleck, and Jörgens 2019; Jörgens, Kolleck, and Saerbeck 2016; Kolleck et al. 2017; Saerbeck et al. 2020; Well et al. 2020; see also Chapter 9). For example, using Twitter data, Kolleck et al. (2017) find that the UNFCCC Secretariat occupied a central and potentially influential position within education-specific communication networks in UNFCCC negotiations from 2009 to 2014. Saerbeck et al. (2020) corroborate this finding with data from an original large-N survey, showing that the climate secretariat was among the five most central organizations during the negotiations leading up to the Paris Agreement. More than other actors, it maintains strong links with a wide range of state and nonstate actors, which allows it to act as a policy broker between different types of actors in global climate governance.

Studying the centrality of IPAs in policy networks and how this centrality relates to the potential influence of IPAs on negotiation processes and outputs requires innovative methods. Against this backdrop, Jörgens, Kolleck, and Saerbeck (2016) argue that SNA a promising method for assessing the political influence of IPAs. Instead of relying on an actor's openly expressed policy preferences or on its reputation for being influential, SNA infers influence from the actor's relative position in issue-specific communication networks (Kolleck 2016). However, descriptive techniques of SNA are only able to assess an actor's potential influence. In order to study whether IPAs are actually willing and able to exploit this potential, inferential techniques of SNA as well as a combination of quantitative SNA with qualitative methods may result in a more accurate picture of secretariat influence and lead to a better/deeper understanding of the causal mechanisms through which it becomes possible. Kolleck et al. (2017), for example, combine SNA with participant observation in their study on the role of the climate secretariat in promoting climate change education. Kolleck (2016), Kolleck, Jörgens, and Well (2017), and Goritz, Jörgens, and Kolleck (2021, 2022) apply current advancements of the inferential techniques of SNA to enable inferential conclusions based on large datasets. Saerbeck et al.

(2020) combine their survey-based SNA with insights from thirty-three semistructured interviews to better understand whether and how the climate secretariat uses its brokerage position to shape issue-specific information flows.

Nevertheless, the centrality and influence of international bureaucracies are not necessarily limited to individual issue areas. Often, they can also be found to operate at the boundary between two or more neighboring policy subdomains. Based on her research on overlap management between international environmental regimes, Jinnah (2012: 108) argues that their "characteristics uniquely position secretariats to manage regime overlap more efficiently and effectively than other actors." "When it comes to coordination of daily, weekly, or even monthly activities between large numbers of actors across two or more international regimes, there is nobody better suited to manage the process than Secretariat staff" (Jinnah 2012: 109). In a similar vein, Jönsson (1986: 42) suggests that at the international level "[b]oundary-role occupants ... are typically found within the secretariat." We therefore expect international bureaucracies to occupy central positions at the intersection of different environmental issue areas. In fact, their centrality and potential for influence may turn out to be even greater if we shift our focus from individual IPAs to networks of bureaucracies operating at different levels of government within a given policy domain.

Against this backdrop, Barbara Saerbeck et al. explore in Chapter 9 whether a global administrative space in environmental governance is emerging that combines the development and strengthening of independent administrative capacities at the international level with the increasing integration of a broad range of governmental and nongovernmental organizations (NGOs) at different levels of government. This administrative space constitutes a complex multilevel and multiactor structure for the management of global environmental policies. Based on an original dataset covering issue-specific cooperation and communication flows between organizations and with regard to the negotiation and implementation of two international environmental conventions, the UNFCCC and the CBD, the authors use SNA to describe and analyze the structure and integration of administrative networks in the environmental field. The exploratory study finds a relatively stable pattern of mutual interaction among the two convention secretariats, other IOs, national and subnational ministries and agencies, research institutes, and NGOs that can be interpreted as an indicator for the emergence of a global environmental administrative space.

### *International Civil Servants*

Not only the organizational and relational aspects of international bureaucracies but also the characteristics of the international civil servants who work within these bureaucracies can affect the role and influence of IPAs (Ege 2020). In the literature, several characteristics of international bureaucrats have been pointed out that may affect an IPA's potential influence.

First, and contrary to its national counterparts, the international civil service is multinational. The staff of IOs or of departments within them are never recruited from just one member country. Even in the case of the directorates general of the European Commission, which are led by nationally appointed commissioners and therefore are sometimes regarded as national domains within the supranational Commission, the civil servants stem from various member states. As a consequence, international civil servants are motivated by departmental, epistemic, and supranational concerns rather than national loyalties (for the European Commission, see Trondal 2006).

Second, international civil servants can be expected to be at least partially driven by professional or normative beliefs. As professionals they are committed to developing and promoting effective solutions to the policy problems they are confronted with. As Michaelowa and Michaelowa in Chapter 7 remind us, "since bureaucrats are not hired at random, but from a community of people who self-selected into this specific field of activity in the first place, they should also be expected to be more dedicated to this field than the average citizen." It is against this backdrop that Barnett and Finnemore (1999: 713) characterized international civil servants as "the 'missionaries' of our time."

Third, various studies suggest that civil servants at all levels of government – from IPAs to independent regulatory agencies to national and subnational bureaucracies – may form domain- or issue-specific epistemic communities that share a set of normative and causal beliefs regarding problem definitions and policy preferences to address these problems. These epistemic beliefs are supranational rather than rooted in notions of national interest. Already in 1971, Jacobson (1971: 780) argued that civil servants operating at different levels of government but within the same issue area develop common sets of interests and priorities. Referring to these epistemic communities as "metabureaucracies," he observed:

The secretariats of international organizations are indeed bureaucracies, but the conference machinery is also composed predominantly of bureaucrats. ... The bureaucrats who make up the conference machinery of international organizations, particularly those operating in technical fields, have interests that are often very closely linked with those of the international secretariat; there is a sectorially shared sense of priorities. Hence the conference machinery does not exercise control over an international secretariat in the same way that, for example, a legislature does.

*Jacobson (1971: 780)*

There seems to be a global administrative space that emerges not only around bureaucratic organizations but also around their permanent staff. Already in 1974, Levi (1974: 51–52) referred to this as "an international political culture," which "is evident in several aspects of international politics. The similarity in the behavior of officials representing their states on the international scene, in the demands they present, in the solutions they suggest, is astonishing. They appear to have lost most of their 'national character.'"

In sum, we expect that in political arenas where civil servants at all levels of government have significant autonomy of action, notions of national interest are less prominent and cooperation is more focused on supranational gains. This would especially be the case in policy domains where a densely populated international administrative space can be observed (see Chapter 9).

## 1.6 Methodological Chances and Challenges of Studying IPA Influence

In methodological terms, a key challenge for IPA researchers is to measure the influence of international bureaucracies against that of other relevant actors. Unlike national and subnational governments, political parties, NGOs, or private sector lobby groups, international secretariats normally refrain from stating their policy preferences in publicly available position papers or manifestos. As a consequence, most of the established methods to empirically infer the influence of political actors – the attributed influence method and the assessment of preference attainment (Betsill and Corell 2008; Dür 2008; Klüver 2013) – are of limited use when focusing on international bureaucracies. New methods for assessing the influence of international bureaucracies that complement and go beyond the traditional combination of interviews and document analysis need to be developed.

Against this backdrop, Chapter 7 by Michaelowa and Michaelowa combines longitudinal data on staff and budget growth with expert interviews, document analysis, and data obtained from CDM databases to infer changes in the climate secretariat's influence on the technical regulation of the CDM mechanism over time. The authors argue that the increased revenue from the CDM both directly and indirectly strengthened the role of the bureaucracy. Conversely, when this revenue decreased, the UNFCCC Secretariat lost part of its autonomous regulatory influence on the CDM and "tried to reorient CDM resources towards support of the Paris Agreement negotiations." In Chapter 3 by Hickmann et al. and Chapter 5 by Hall, the authors also take a longitudinal stance as they analyze the growing autonomy and influence of international environmental secretariats and their executive leadership over time. In Chapter 6, Lynn Wagner and Pamela Chasek systematically study change in secretariat financing over time in order to shed light on the ways in which states attempt to gain "control over the focus of activity and level of ambition that secretariats can undertake." Wagner and Chasek's account complements Chapters 5 and 7 in that it zooms in on the states' side of the principal–agent relationship, that is, on how the parties to international environmental conventions attempt to exert control over IPAs such as the convention secretariats, and contrasts this view with the agent perspective expressed in Chapters 5 and 8. The chapter explores the negotiation dynamics and budget decisions regarding three UN conventions – the UNFCCC, the CBD, and the United Nations Convention to Combat Desertification – as well as the two multilateral scientific bodies, the International Panel on Climate Change and the Intergovernmental Science-Policy Platform on Biodiversity and Ecosystem Services. Contrasting the finding of some of the previous chapters that international secretariats are to a certain extent able to circumvent control by their state principals, Wagner and Chasek show that states continue to oversee and control their bureaucratic agents beyond the original delegation contract. Recurring program and budget negotiations are found to be a key mechanism that enables states to react to tendencies of secretariats to increase their autonomy and their subsequent influence on the policies of IOs.

## 1.7 Conclusion

This book unites a variety of innovative contributions, new conceptual approaches, and empirical findings by some of the most renowned authors in this field of study. It offers a comprehensive resource for the study of IPAs in global environmental politics. Conceptually, it is thought to provide both theoretical and methodological perspectives as well as cutting-edge empirical studies, each with clear reference to the policymaking role of international environmental bureaucracies. Methodologically, it uses different quantitative and qualitative techniques to measure the influence of IPAs to an empirical test and provides a solid overview on the chances and challenges of research methodologies in an increasingly relevant research field. Empirically, it gives an overview of pioneering case study research on international environmental bureaucracies across different issue areas in environmental policymaking. The book is thus aimed both at scientists from the fields of global environmental policy and international administration and at practitioners who are directly confronted with the challenges of these new forms of transnational influence.

Hence, the book will appeal to researchers in the field of global environmental politics and also to practitioners working for international administrations, IOs, national delegations, or civil society organizations. For practitioners, the book's subject is relevant and timely for at least three reasons: First, members of national delegations at multilateral negotiations arguably have a vital interest in understanding how different actors strive for influence and control during and in between negotiations, in order to determine the relationship to other actors that is beneficial for them. Therefore, understanding the role of IPAs as actors in global environmental politics, their interests, motives, and strategies, can be a strategic advantage. While the principal–agent relationship between national delegations and IPAs can be regarded as one prominent, widely shared point of reference for practitioners (in the present example in their role as parties, i.e., principals), in this book we argue that understanding IPAs as partially autonomous actors with their own interests and motivations will provide valuable insights for (state) practitioners regarding their own strategic interaction with IPAs and other actors during and between multilateral negotiations. Parties can also conceive of IPAs as potential partners, rather than agents or instruments. They can thus seek productive collaboration with them, use the relationship with them strategically, or tap into IPAs' unique expertise about policies and processes. As argued in Chapter 4 by Well et al., IPAs try to present themselves this way when they act as "attention-seeking bureaucracies." Thinking beyond the principal–agent roles can have a practical advantage for practitioners and in this sense be liberating and productive. This is also true for nonstate practitioners because nonstate actors in multilateral negotiations increasingly have a fingerprint on multilateral processes and may use their relationships to IPAs to further their goals. Chapters 2 to 4 work with concepts that emphasize this perspective. On the other side, Chapters 6 and 7 show very clearly what merit the principal–agent approach continues to have by providing a detailed analysis of how parties exert or gradually lose control over IPAs owing to their (in)ability to control secretariat financing.

Second, IPAs act as brokers and strategically connect negotiation parties as well as nonparty and party stakeholders with one another. This perspective is empirically underpinned in Chapter 9. It invites practitioners to understand the actor network they work in as an emerging global administrative space, in which the connection to IPAs can be of strategic importance for the impact one organization can have in the policy network. Conceiving of IPAs as brokers of information and policy ideas in international environmental politics and positioning oneself vis-à-vis these actors can be a powerful tool.

Finally, state and nonstate practitioners may be interested in understanding how democratically legitimate certain practices observed among IPAs are (Chapter 8). This allows questions to be answered about the normative desirability of IPAs' tendency to become actors in their own right. Understanding these aspects is certainly valuable for informing and justifying state and nonstate policies and choices in an evolving multilateral setting, for example, with regard to institutional design, development, or reform concerning existing or emerging policy issues.

We hope that with this book we can stimulate debate on the influence of international secretariats in global environmental governance, inspire and inform practitioners in the field, advance knowledge, and encourage further studies in a dynamic and increasingly relevant field of research.

## References

Abbott, K. W. and Snidal, D. (1998). Why States Act through Formal International Organizations, *Journal of Conflict Resolution* 42 (1): 3–32.

Abbott, K. W. and Snidal, D. (2010). International Regulation without International Government: Improving IO Performance through Orchestration, *Review of International Organizations* 5 (3): 315–344.

Abbott, K. W., Genschel, P., Snidal, D., and Zangl, B. (eds.) (2015). *International Organizations as Orchestrators*, Cambridge: Cambridge University Press.

Bäckstrand, K. and Kuyper, J. W. (2017). The Democratic Legitimacy of Orchestration: The UNFCCC, Non-state Actors, and Transnational Climate Governance, *Environmental Politics* 26 (4): 764–788.

Barnett, M. and Coleman, L. (2005). Designing Police: Interpol and the Study of Change in International Organizations, *International Studies Quarterly* 49 (4): 593–620.

Barnett, M. and Finnemore, M. (1999). The Politics, Power, and Pathologies of International Organizations, *International Organization* 53 (4): 699–732.

Barnett, M. and Finnemore, M. (2004). *Rules for the World: International Organizations in Global Politics*, Ithaca, NY: Cornell University Press.

Bauer, M. W. and Ege, J. (2016). Bureaucratic Autonomy of International Organizations' Secretariats, *Journal of European Public Policy* 23 (7): 1019–1037.

Bauer, M. W. and Ege, J. (2017). A Matter of Will and Action: The Bureaucratic Autonomy of International Public Administrations. In M. W. Bauer, C. Knill, and S. Eckhard (eds.), *International Bureaucracy: Challenges and Lessons for Public Administration Research*, Basingstoke: Palgrave Macmillan, 13–41.

Bauer, M. W., Knill, C., and Eckhard, S. (eds.) (2017a). *International Bureaucracy: Challenges and Lessons for Public Administration Research*, Basingstoke: Palgrave Macmillan.

Bauer, M. W., Knill, C., and Eckhard, S. (2017b). International Public Administration: A New Type of Bureaucracy? Lessons and Challenges for Public Administration Research. In M. W. Bauer, C. Knill and S. Eckhard (eds.), *International Bureaucracy: Challenges and Lessons for Public Administration Research*, Basingstoke: Palgrave Macmillan, 179–198.

Bauer, M. W., Eckhard, S., Ege, J., and Knill, C. (2017). A Public Administration Perspective on International Organizations. In M. W. Bauer, C. Knill, and S. Eckhard (eds.), *International Bureaucracy: Challenges and Lessons for Public Administration Research*, Basingstoke: Palgrave Macmillan, 1–12.

Bauer, S. (2006). Does Bureaucracy Really Matter? The Authority of Intergovernmental Treaty Secretariats in Global Environmental Politics, *Global Environmental Politics* 6 (1): 23–49.

Bauer, S. (2009). The Desertification Secretariat: A Castle Made of Sand. In F. Biermann and B. Siebenhüner (eds.), *Managers of Global Change: The Influence of International Environmental Bureaucracies*, Cambridge, MA: MIT Press, 293–317.

Bauer, S. and Weinlich, S. (2011). International Bureaucracies: Organizing World Politics. In B. Reinalda (ed.), *The Ashgate Research Companion to Non-state Actors*, Farnham: Ashgate, 251–262.

Bayerlein, L., Knill, C., and Steinebach, Y. (2020). *A Matter of Style? Organizational Agency in Global Public Policy, Cambridge Studies in Comparative Public Policy*, Cambridge: Cambridge University Press.

Beach, D. (2008). The Facilitator of Efficient Negotiations in the Council: The Impact of the Council Secretariat. In D. Naurin and H. Wallace (eds.), *Unveiling the Council of the European Union: Games Governments Play in Brussels*, Basingstoke: Palgrave Macmillan, 219–237.

Benz, A., Corcaci, A., and Doser, J. W. (2017). Multilevel Administration in International and National Contexts. In M. W. Bauer, C. Knill, and S. Eckhard (eds.), *International Bureaucracy: Challenges and Lessons for Public Administration Research*, Basingstoke: Palgrave Macmillan, 151–178.

Betsill, M. M. and Corell, E. (2008). Analytical Framework: Assessing the Influence of NGO Diplomats. In M. M. Betsill and E. Corell (eds.), *NGO Diplomacy: The Influence of Nongovernmental Organizations in International Environmental Negotiations*, Cambridge, MA: MIT Press, 19–42.

Bhattacharya, A. K. (1976). The Influence of the International Secretariat: UNCTAD and Generalized Tariff Preferences, *International Organization* 30 (1): 75–90.

Biermann, F. and Siebenhüner, B. (2009a). The Influence of International Bureaucracies in World Politics: Findings from the MANUS Research Program. In F. Biermann and B. Siebenhüner (eds.), *Managers of Global Change: The Influence of International Environmental Bureaucracies*, Cambridge, MA: MIT Press, 319–349.

Biermann, F. and Siebenhüner, B. (eds.) (2009b). *Managers of Global Change: The Influence of International Environmental Bureaucracies*, Cambridge, MA: MIT Press.

Biermann, F. and Siebenhüner, B. (2009c). The Role and Relevance of International Bureaucracies: Setting the Stage. In F. Biermann and B. Siebenhüner (eds.), *Managers of Global Change: The Influence of International Environmental Bureaucracies*, Cambridge, MA: MIT Press, 1–14.

Biermann, F., Siebenhüner, B., Bauer, S., et al. (2009). Studying the Influence of International Bureaucracies: A Conceptual Framework. In F. Biermann and B. Siebenhüner (eds.), *Managers of Global Change: The Influence of International Environmental Bureaucracies*, Cambridge, MA: MIT Press, 37–74.

Bohne, E. (2010). *The World Trade Organization: Institutional Development and Reform*, Basingstoke: Palgrave Macmillan.

Busch, P.-O. (2009a). The Climate Secretariat: Making a Living in a Straitjacket. In F. Biermann and B. Siebenhüner (eds.), *Managers of Global Change: The Influence of International Environmental Bureaucracies*, Cambridge, MA: MIT Press, 245–264.

Busch, P.-O. (2009b). The OECD Environment Directorate: The Art of Persuasion and Its Limitations. In F. Biermann and B. Siebenhüner (eds.), *Managers of Global Change: The Influence of International Environmental Bureaucracies*, Cambridge, MA: MIT Press, 75–99.

Busch, P.-O. (2014). The Independent Influence of International Public Administrations: Contours and Future Directions of an Emerging Research Strand. In S. Kim, S. Ashley, and W. H. Lambright (eds.), *Public Administration in the Context of Global Governance*, Cheltenham: Edward Elgar, 45–62.

Busch, P.-O. and Liese, A. (2017). The Authority of International Public Administrations. In M. W. Bauer, C. Knill, and Eckhard, S. (eds.), *International Bureaucracy: Challenges and Lessons for Public Administration Research*, Basingstoke: Palgrave Macmillan, 97–122.

Busch, P.-O., Feil, H., Heinzel, M., et al. (2021). Policy Recommendations of International Bureaucracies: The Importance of Country-Specificity, *International Review of Administrative Sciences* 87 (4): 775–793.

Chan, S., van Asselt, H., Hale, T., et al. (2015). Reinvigorating International Climate Policy: A Comprehensive Framework for Effective Nonstate Action, *Global Policy* 6 (4): 466–473.

Churchill, R. R. and Ulfstein, G. (2000). Autonomous Institutional Arrangements in Multilateral Environmental Agreements: A Little-Noticed Phenomenon in International Law, *American Journal of International Law* 94 (4): 623–659.

Da Conceição, E. (2010). Who Controls Whom? Dynamics of Power Delegation and Agency Losses in EU Trade Politics, *Journal of Common Market Studies* 48 (4): 1107–1126.

Dehousse, R. (2008). Delegation of Powers in the European Union: The Need for a Multi-principals Model, *West European Politics* 31 (4): 789–805.

Depledge, J. (2007). A Special Relationship: Chairpersons and the Secretariat in the Climate Change Negotiations, *Global Environmental Politics* 7 (1): 45–68.

Derlien, H.-U., Böhme, D., and Heindl, M. (2011). *Bürokratietheorie. Einführung in eine Theorie der Verwaltung*, Wiesbaden: VS Verlag für Sozialwissenschaften.

Dijkstra, H. (2010). Explaining Variation in the Role of the EU Council Secretariat in First and Second Pillar Policy-Making, *Journal of European Public Policy* 17 (4): 527–544.

Dijkstra, H. (2017). Collusion in International Organizations: How States Benefit from the Authority of Secretariats, *Global Governance: A Review of Multilateralism and International Organizations* 23 (4): 601–619.

Dimitrov, R. S. (2010). Inside UN Climate Change Negotiations: The Copenhagen Conference, *Review of Policy Research* 27 (6): 795–821.

Dingwerth, K. and Jörgens, H. (2015). Environmental Risks and the Changing Interface of Domestic and International Governance. In S. Leibfried, F. Nullmeier, E. Huber, et al. (eds.), *The Oxford Handbook of Transformations of the State*, Oxford: Oxford University Press, 338–354.

Dür, A. (2008). Interest Groups in the European Union: How Powerful Are They?, *West European Politics* 31 (6): 1212–1230.

Eckhard, S. and Ege, J. (2016). International Bureaucracies and Their Influence on Policy-Making: A Review of Empirical Evidence, *Journal of European Public Policy* 23 (7): 960–978.

Ege, J. (2020). What International Bureaucrats (Really) Want, *Global Governance: A Review of Multilateralism and International Organizations* 26 (4): 577–600.

Ellis, D. C. (2010). The Organizational Turn in International Organization Theory, *Journal of International Organizations Studies* 1 (1): 11–28.

Fernandez, R. M. and Gould, R. V. (1994). A Dilemma of State Power: Brokerage and Influence in the National Health Policy Domain, *American Journal of Sociology* 99 (6): 1455–1491.

Finkelstein, L. S. (1974). International Organizations and Change: The Past as Prologue, *International Studies Quarterly* 18 (4): 485–520.

Fleischer, J. and Reiners, N. (2021). Connecting International Relations and Public Administration: Toward a Joint Research Agenda for the Study of International Bureaucracy, *International Studies Review* 23 (4): 1230–1247.

Gehring, T. (2012). International Environmental Regimes as Decision Machines. In P. Dauvergne (ed.), *Handbook of Global Environmental Politics*, 2nd ed., Cheltenham: Edward Elgar, 51–63.

Goritz, A., Jörgens, H., and Kolleck, N. (2021). Interconnected Bureaucracies? Comparing Online and Offline Networks during Global Climate Negotiations, *International Review of Administrative Sciences* 87 (4). DOI: 10.1177/00208523211022823.

Goritz, A., Jörgens, H., and Kolleck, N. (2022). A Matter of Information: The Influence of International Bureaucracies in Global Climate Governance Networks, *Social Networks*. DOI: 10.1016/j.socnet.2022.02.009.

Goritz, A., Kolleck, N., and Jörgens, H. (2019). Education for Sustainable Development and Climate Change Education: The Potential of Social Network Analysis Based on Twitter Data, *Sustainability* 11 (19): 5499.

Goritz, A., Schuster, J., Jörgens, H., and Kolleck, N. (2020). International Public Administrations on Twitter: A Comparison of Digital Authority in Global Climate Policy, *Journal of Comparative Policy Analysis: Research and Practice* 24 (11): 1–25.

Gutner, T. L. (2005). World Bank Environmental Reform: Revisiting Lessons from Agency Theory, *International Organization* 59 (3): 773–783.

Hale, T., and Roger, C. (2014). Orchestration and Transnational Climate Governance, *Review of International Organizations* 9 (1): 59–82.

Hawkins, D. and Jacoby, W. (2006). How Agents Matter. In D. Hawkins, D. A. Lake, D. L. Nielson, and M. J. Tierney (eds.), *Delegation and Agency in International Organizations*, Cambridge: Cambridge University Press, 199–228.

Hawkins, D., Lake, D. A., Nielson, D. L., and Tierney, M. J. (eds.) (2006a). *Delegation and Agency in International Organizations*, Cambridge: Cambridge University Press.

Hawkins, D., Lake, D. A., Nielson, D. L., and Tierney, M. J. (2006b). Delegation under Anarchy: States, International Organizations, and Principal-Agent Theory. In D. Hawkins, D. A. Lake, D. L. Nielson, and M. J. Tierney (eds.), *Delegation and Agency in International Organizations*. Cambridge: Cambridge University Press, 3–38.

Herold, J., Liese, A., Busch, P.-O., and Feil, H. (2021). Why National Ministries Consider the Policy Advice of International Bureaucracies: Survey Evidence from 106 Countries, *International Studies Quarterly* 65 (3): 669–682.

Hickmann, T. (2019). Authority in World Politics, *Oxford Research Encyclopedia of International Studies*. DOI: 10.1093/acrefore/9780190846626.013.502.

Jacobson, H. K. (1971). Technological Developments, Organizational Capabilities, and Values, *International Organization* 25 (4): 776–783.

Jinnah, S. (2010). Overlap Management in the World Trade Organization: Secretariat Influence on Trade-Environment Politics, *Global Environmental Politics* 10 (2): 54–79.

Jinnah, S. (2011). Marketing Linkages: Secretariat Governance of the Climate-Biodiversity Interface, *Global Environmental Politics* 11 (3): 23–43.

Jinnah, S. (2012). Singing the Unsung: Secretariats in Global Environmental Politics. In P. S. Chasek and L. M. Wagner (eds.), *The Roads from Rio: Lessons Learned from Twenty Years of Multilateral Environmental Negotiations*, London: Routledge, 107–126.

Jinnah, S. (2014). *Post-Treaty Politics: Secretariat Influence in Global Environmental Governance*, Cambridge, MA: MIT Press.

Johnson, T. (2013). Institutional Design and Bureaucrats' Impact on Political Control, *Journal of Politics* 75 (1): 183–197.

Johnson, T. (2014). *Organizational Progeny: Why Governments Are Losing Control over the Proliferating Structures of Global Governance*, Oxford: Oxford University Press.

Johnson, T. and Urpelainen, J. (2014). International Bureaucrats and the Formation of Intergovernmental Organizations: Institutional Design Discretion Sweetens the Pot, *International Organization* 68 (1): 177–209.

Johnstone, I. (2007). The Secretary-General as Norm Entrepreneur. In S. Chesterman (ed.), *Secretary or General? The UN Secretary-General in World Politics*, Cambridge: Cambridge University Press, 123–138.

Jönsson, C. (1986). Interorganization Theory and International Relations, *International Studies Quarterly* 30 (1): 39–57.

Jörgens, H., Kolleck, N., and Saerbeck, B. (2016). Exploring the Hidden Influence of International Treaty Secretariats: Using Social Network Analysis to Analyse the Twitter Debate on the "Lima Work Programme on Gender," *Journal of European Public Policy* 23 (7): 979–998.

Jörgens, H., Kolleck, N., Saerbeck, B., and Well, M. (2017). Orchestrating (Bio-)Diversity: The Secretariat of the Convention of Biological Diversity as an Attention-Seeking Bureaucracy. In M. W. Bauer, C. Knill, and S. Eckhard (eds.), *International Bureaucracy: Challenges and Lessons for Public Administration Research*, Basingstoke: Palgrave Macmillan, 73–95.

Keohane, R. O. (1969). Institutionalization in the United Nations General Assembly, *International Organization* 23 (4): 859–896.

Keohane, R. O. (1984). *After Hegemony: Cooperation and Discord in the World Political Economy*, Princeton: Princeton University Press.

Keohane, R. O. and Nye, J. S. (1971). Preface, *International Organization* 25 (3): v–vi.

Keohane, R. O. and Nye, J. S. (1974). Transgovernmental Relations and International Organizations, *World Politics* 27 (1): 39–62.

Keohane, R. O., and Victor, D. G. (2011). The Regime Complex for Climate Change, *Perspectives on Politics* 9 (1): 7–23.

Klüver, H. (2013). *Lobbying in the European Union: Interest Groups, Lobbying Coalitions, and Policy Change*, Oxford: Oxford University Press.

Knill, C. and Bauer, M. W. (2016). Policy-Making by International Public Administrations: Concepts, Causes and Consequences, *Journal of European Public Policy* 23 (7): 949–959.

Knill, C., Bayerlein, L., Enkler, J., and Grohs, S. (2018). Bureaucratic Influence and Administrative Styles in International Organizations, *The Review of International Organizations* 14 (1): 83–106.

Knill, C., Enkler, J., Schmidt, S., Eckhard, S., and Grohs, S. (2017). Administrative Styles of International Organizations: Can We Find Them, Do They Matter? In M. W. Bauer, C. Knill, and S. Eckhard (eds.), *International Bureaucracy: Challenges and Lessons for Public Administration Research*, Basingstoke: Palgrave Macmillan, 43–71.

Kolleck, N. (2016). Uncovering Influence through Social Network Analysis: The Role of Schools in Education for Sustainable Development, *Journal of Education Policy* 31 (3): 308–330.

Kolleck, N., Jörgens, H., and Well, M. (2017). Levels of Governance in Policy Innovation Cycles in Community Education: The Cases of Education for Sustainable Development and Climate Change Education, *Sustainability* 9 (11): 1966.

Kolleck, N., Well, M., Sperzel, S., and Jörgens, H. (2017). The Power of Social Networks: How the UNFCCC Secretariat Creates Momentum for Climate Education, *Global Environmental Politics* 17 (4): 106–126.

Lenz, T., Bezuijen, J., Hooghe, L., and Marks, G. (2015). Patterns of International Organization: Task Specific vs. General Purpose. In E. da Conceição-Heldt, M. Koch, and A. Liese (eds.), *Internationale Organisationen: Autonomie, Politisierung, interorganisationale Beziehungen und Wandel, Politische Vierteljahresschrift* (Sonderheft 49), Baden-Baden: Nomos, 131–156.

Levi, W. (1974). *International Politics: Foundations of the System*, Minneapolis: University of Minnesota Press.

Levinson, N. S. and Marzouki, M. (2016). International Organizations and Global Internet Governance: Interorganizational Architecture. In F. Musiani, D. L. Cogburn, L. DeNardis, and N. S. Levinson (eds.), *The Turn to Infrastructure in Internet Governance*, Basingstoke: Palgrave Macmillan, 47–71.

Liese, A. and Weinlich, S. (2006). Die Rolle von Verwaltungsstäben internationaler Organisationen: Lücken, Tücken und Konturen eines (neuen) Forschungsgebiets. In J. Bogumil, W. Jann, and F. Nullmeier (eds.), *Politik und Verwaltung, Politische Vierteljahresschrift* (Sonderheft 37), Wiesbaden: VS Verlag für Sozialwissenschaften, 491–524.

Martin, L. L. and Simmons, B. A. (2012). International Organizations and Institutions. In W. Carlsnaes, T. Risse, and B. A. Simmons (eds.), *Handbook of International Relations*, 2nd ed., London: SAGE Publications, 326–351.

Mathiason, J. (2007). *Invisible Governance: International Secretariats in Global Politics*, Bloomfield: Kumarian Press.

Mederake, L., Saerbeck, B., Goritz, A., et al. (2021). Cultivated Ties and Strategic Communication: Do International Environmental Secretariats Tailor Information to Increase their Bureaucratic Reputation? *International Environmental Agreements: Politics, Law and Economics* (published online). DOI: 10.1007/s10784-021-09554-3.

Ness, G. D. and Brechin, S. R. (1988). Bridging the Gap: International Organizations as Organizations, *International Organization* 42 (2): 245–273.

Newell, P., Pattberg, P., and Schroeder, H. (2012). Multiactor Governance and the Environment, *Annual Review of Environment and Resources* 37 (1): 365–387.

Nielson, D. L. and Tierney, M. J. (2003). Delegation to International Organizations: Agency Theory and World Bank Environmental Reform, *International Organization* 57 (2): 241–276.

Patz, R. and Goetz, K. H. (2017). Changing Budgeting Administration in International Organizations: Budgetary Pressures, Complex Principals and Administrative Leadership. In M. W. Bauer, C. Knill, and S. Eckhard (eds.), *International Bureaucracy: Challenges and Lessons for Public Administration Research*, Basingstoke: Palgrave Macmillan, 123–150.

Pevehouse, J. C., Nordstrom, T., and Warnke, K. (2005). International Governmental Organizations. In P. F. Diehl (ed.), *The Politics of Global Governance: International Organizations in an Interdependent World*, 3rd ed., Boulder, CO: Lynne Rienner, 9–24.

Reinalda, B. and Verbeek, B. (1998). Autonomous Policy Making by International Organizations: Purpose, Outline, and Results. In B. Reinalda and B. Verbeek (eds.), *Autonomous Policy Making by International Organizations*, London: Routledge, 1–8.

Saerbeck, B., Well, M., Jörgens, H., Goritz, A., and Kolleck, N. (2020). Brokering Climate Action: The UNFCCC Secretariat between Parties and Nonparty Stakeholders, *Global Environmental Politics* 20 (2): 105–127.

Sandford, R. (1994). International Environmental Treaty Secretariats: Stage-Hands or Actors? In H. O. Bergesen and G. Parmann (eds.), *Green Globe Yearbook of International Cooperation on Environment and Development 1994*, Oxford: Oxford University Press, 17–29.

Schmitter, P. C. (1971). The "Organizational Development" of International Organizations, *International Organization* 25 (4): 917–937.

Shanks, C., Jacobson, H. K., and Kaplan, J. H. (1996). Inertia and Change in the Constellation of International Governmental Organizations, 1981–1992, *International Organization* 50 (4): 593–627.

Siebenhüner, B. (2007). Administrator of Global Biodiversity: The Secretariat of the Convention on Biological Diversity, *Biodiversity and Conservation* 16 (1): 259–274.

Siebenhüner, B. (2009). The Biodiversity Secretariat: Lean Shark in Troubled Waters. In F. Biermann and B. Siebenhüner (eds.), *Managers of Global Change: The Influence of International Environmental Bureaucracies*, Cambridge, MA: MIT Press, 265–291.

Siotis, J. (1965). The Secretariat of the United Nations Economic Commission for Europe and European Economic Integration: The First Ten Years, *International Organization* 19 (2): 177–202.

Tallberg, J. (2010). The Power of the Chair: Formal Leadership in International Cooperation, *International Studies Quarterly* 54 (1): 241–265.

Trondal, J. (2006). Governing at the Frontier of the European Commission: The Case of Seconded National Officials, *West European Politics* 29 (1): 147–160.

Trondal, J. (2017). A Research Agenda on International Public Administration. In J. Trondal (ed.), *The Rise of Common Political Order: Institutions, Public Administration and Transnational Space*, Cheltenham: Edward Elgar, 35–48.

Vaubel, R. (2006). Principal-Agent Problems in International Organizations, *Review of International Organizations* 1 (2): 125–138.

Verhoest, K., Roness, P. G., Verschuere, B., Rubecksen, K., and MacCarthaigh, M. (2010). *Autonomy and Control of State Agencies: Comparing States and Agencies*, Basingstoke: Palgrave Macmillan.

Wagner, L. M. and Mwangi, M. (2010). Be Careful What You Compromise For: Postagreement Negotiations within the UN Desertification Convention, *International Negotiation* 15 (3): 439–458.

Weinlich, S. (2014). *The UN Secretariat's Influence on the Evolution of Peacekeeping, Transformations of the State*, Basingstoke: Palgrave Macmillan.

Well, M., Saerbeck, B., Jörgens, H., and Kolleck, N. (2020). Between Mandate and Motivation: Bureaucratic Behavior in Global Climate Governance, *Global Governance: A Review of Multilateralism and International Organizations* 26 (1): 99–120.

Wit, D. de, Ostovar, A. L., Bauer, S., and Jinnah, S. (2020). International Bureaucracies. In F. Biermann and R. E. Kim (eds.), *Architectures of Earth System Governance: Institutional Complexity and Structural Transformation*, Cambridge: Cambridge University Press, 57–74.

Zelli, F., and Asselt, H. van (2013). Introduction: The Institutional Fragmentation of Global Environmental Governance: Causes, Consequences, and Responses, *Global Environmental Politics* 13 (3): 1–13.

# 2

# Means of Bureaucratic Influence

*The Interplay between Formal Autonomy and Informal Styles in International Bureaucracies*

MICHAEL W. BAUER, STEFFEN ECKHARD, JÖRN EGE, AND CHRISTOPH KNILL

## 2.1 Introduction

International public administrations (IPAs), that is, the secretariats of international governmental organizations (IGOs) that constitute the international counterparts to administrative bodies at national and subnational levels, have attracted considerable scholarly attention in recent years (Barnett and Finnemore 2004; Bauer et al. 2017; Ege and Bauer 2013; Knill and Bauer 2016; Liese and Weinlich 2006; Thorvaldsdottir, Patz, and Eckhard 2021). While several studies ascribe an influential role to IPAs in a variety of policy fields (Biermann and Siebenhüner 2009; Ege, Bauer, and Wagner 2021; Nay 2012; Reinalda and Verbeek 2004; Skovgaard 2017; Stone and Ladi 2015; Stone and Moloney 2019), the questions of to what degree and under which precise conditions these bodies influence the making and application of international public policies are still vividly debated (see Eckhard and Ege 2016; Ege, Bauer, and Wagner 2020). Given the increasing significance of global environmental challenges as discussed in this book, the question of independent influence is particularly relevant for international environmental bureaucracies (see Chapter 1). Instead of studying the secretariats of multilateral environmental conventions, however, we want to focus on the question of bureaucratic influence of larger and more institutionalized international bureaucracies, which nevertheless play an important role in global environmental governance (see Chapter 9). Comparing the administrations of the Food and Agriculture Organization (FAO), the United Nations Educational, Scientific and Cultural Organization (UNESCO), and the Organisation for Economic Co-operation and Development (OECD), which are involved in environmental governance with IPAs in other sectors, gives us the opportunity to determine if environmentally active administrations are characterized by common empirical configurations of style and autonomy and thus can be expected to exhibit a particular policy influence potential.

From a public administration and organizational theory perspective, the role and impact of specific administrative characteristics of international bureaucracies regarding their financial and personnel resources, their competences and expertise, and their specific organizational routines and cultures are of particular interest in the context of this debate (Bauer et al. 2017). In this chapter, we hope to add to this debate about potential bureaucratic influence on policymaking beyond the nation-state in conceptual, theoretical, and empirical terms. When speaking about influence, we depart from the "having an effect" definition prominently introduced by Biermann et al. (2009: 41) who defined IPA influence as "the sum of all effects observable for, and attributable to, an international bureaucracy" (see also Liese and Weinlich 2006: 504; Weinlich 2014: 60–61). Yet for our analytical purpose we modify this definition insofar as we consider IPAs' influence potentials rather than trying to factually distil the degree of administrative influence on a given policy adopted by an IGO (see Bayerlein, Knill, and Steinebach 2020; Knill et al. 2019). Conceptually, we distinguish between two sources of potential bureaucratic influence, namely formal structural autonomy enjoyed by IPAs and informal behavioral routines as they become apparent in different administrative styles (Davies 1967; Hooghe et al. 2017; Knill 2001; Knill and Grohs 2015; Lall 2017; Simon 1997; Wilson 1989). Structural autonomy and administrative styles are two important aspects (but certainly not the only ones) within the intensively debated explanatory programs with respect to bureaucratic influence: formal administrative structures and informal administrative behavior.

Formal autonomy captures the extent to which an IPA is granted formal competencies and resources to develop and implement public policies. Even though the autonomy concept used here goes beyond formal delegation by also capturing the administrative capacity to develop autonomous preferences (see Bauer, da Conceição-Heldt, and Ege 2015), its operationalization relies on formal organizational characteristics. In this context, researchers typically refer to principal–agent models and highlight the structural relationship between the IPA and its political principals, the member states, expressed in terms of the formal powers and resources member states surrender to the IPA and the control functions they install (Abbott and Snidal 1998; Hawkins et al. 2006a; Hooghe and Marks 2015; Jankauskas 20222; McCubbins, Noll, and Weingast 1989; Stone 2011). In particular, the literature on the rational design of IGOs would expect a higher potential for bureaucratic influence, the higher the levels of formal autonomy of IPAs rise (see, e.g., Ege et al. 2023; Haftel and Thompson 2006; Johnson 2013; Koremenos, Lipson, and Snidal 2001).

Administrative styles, by contrast, capture informal organizational routines that reflect an IPA's institutionalized orientation both in functional terms (policy effectiveness) and in positional terms (institutional consolidation) (Bayerlein, Knill, and Steinebach 2020). Depending on the prevalence of these orientations, IPAs can be conceived as either servant-oriented (trying to read their mission from the lips of their political masters) or entrepreneurial (actively trying to independently push the policymaking activities of their organization in certain directions). In other words, on a continuum between servant-oriented and entrepreneurial style IPAs, one would expect that the more entrepreneurial an IPA is, the more influential it becomes (Knill et al. 2019; Nay 2012; Oksamytna 2018).

We thus take the presumed relationship between autonomy and style, on the one side, and bureaucratic influence, on the other, as our point of departure. However, our prime aim is not to empirically measure and substantiate this relationship but rather to study in more detail how administrative autonomy and administrative style relate to each other in real-world IPAs. This is relevant because with respect to the IPA's formal capacities and informal routines debates have evolved rather isolated from each other. If a systematic theory of the IPA's policy influence is the objective, and if informal and formal administrative patterns are of such importance, as many researchers in the field claim, then the question of how these two bureaucratic dimensions relate to each other in the international sphere is of central analytical interest. In theoretical terms, we therefore want to shed light on the relationship between the formal and the informal sources of bureaucratic influence. We illustrate our theoretical considerations with an empirical assessment of these configurations for nine IGO secretariats operating in different policy fields.

Although there are no IPAs with exclusive environmental policy responsibilities in our sample, our approach is particularly relevant for the study of more specialized environmental bureaucracies such as the secretariats of multilateral conventions. Curiously, attempts to measure the formal autonomy and identify the administrative styles of environmental bureaucracies are rare or even nonexistent in the literature (but see Biermann and Siebenhüner 2009; Widerberg and van Laerhoven 2014). Probably the most systematic attempt to explain influence of environmental bureaucracies with formal factors relating to the "polity" of these organizations was made by Biermann et al. (2009). While the formal mandate, rules, and so on are mentioned in this seminal work, the organization's autonomy is not explicitly defined and operationalized as an explanatory variable. Similarly, this work did not explicitly use the concept of administrative styles, although "people and procedures," including factors such as organizational culture and leadership style, were important variables.

Considering the scarcity of autonomy and style-focused research with respect to international bureaucracies, we believe that the literature on the role and influence of environmental bureaucracies could benefit greatly from adopting the approach presented in this chapter. This seems particularly relevant, first, because of the importance of normative beliefs in the environmental field, which makes a focus on the informal behavior of international civil servants beyond a narrow focus on executive leadership fruitful, and, second, because of the contested nature of costly environmental policies, such as decarbonization strategies, which makes a restriction of formal autonomy of specialized environmental bureaucracies by their principals very likely. In this imaginable context of restricted autonomy combined with deeply rooted normative preferences of IPA staff, our approach can provide an important analytical tool for further research.

Our findings display a variety of configurations. As we will show, there is no clear and dominant pattern in which formal autonomy and administrative styles are linked. A strong and autonomous formal status does not automatically go together with entrepreneurial administrative practices. This is especially visible when looking at the administration of the FAO, UNESCO, and the OECD, which are also active in addressing environmental issues. At the same time, weak autonomy does not necessarily imply that administrative styles reflect a servant type. By shedding light on the complex interactions between formal and informal bureaucratic features, our insights have important implications for the design of accountability mechanisms in view of optimizing bureaucratic control in the international sphere.

The remainder of this chapter is structured as follows: We first present our concepts in more detail to assess formal and informal sources of bureaucratic influence (Section 2.2). We then turn to the theoretical discussion of the relationship between autonomy and administrative styles (Section 2.3). In Section 2.4, we empirically assess different configurations of bureaucratic influence sources within the different IGOs under study. On the basis of our empirical data, we demonstrate how the two concepts link empirically and discuss the relevance and consequence of the emerging patterns – with a particular focus on their potential impact upon policymaking beyond the nation-state.

## 2.2 Conceptualizing and Measuring Sources of IPAs' Bureaucratic Influence

There are many conceivable ways to conceptualize and ultimately measure the influence of international bureaucracies. We do not claim exclusivity for the approach we develop here. We do, however, contend that if the internal characteristics of IPAs are put into focus, then formal as well as informal aspects need to be systematically considered. Furthermore, we see a twofold gap in current research in this area: On the one side, disciplined measurement strategies of both formal and informal concepts are often neglected; on the other side, no attempt is made to investigate whether there is a systematic relationship between formal and informal bureaucratic features – and how these relationships may play out in practice. It is against this background that the following heuristic and analytical suggestions are made.

To capture formal sources of bureaucratic influence, we rely on the concept of structural autonomy (Bauer and Ege 2016a; Ege 2017). The informal potential of bureaucratic influence, by contrast, is assessed on the basis of administrative styles developed by Knill, Eckhard, and Grohs (2016; see Bayerlein, Knill, and Steinebach 2020). With regard to the formality–informality distinction, the difference between the two concepts is visible not only in their conceptualization but also in their operationalization. While the measurement of autonomy relies on formal characteristics, administrative styles are measured based on administrative self-perceptions by means of semistructured interviews with IPA staff members.

### *Structural Bureaucratic Autonomy*

The concept of bureaucratic autonomy is primarily used in the comparative study of regulatory and executive agencies (see Verhoest et al. 2004). Based on the observation that autonomy "means, above all, to be able to translate one's own preferences into authoritative actions" (Maggetti and Verhoest 2014: 239), the concept can also be used to study the structural features of international administrations. To this end, we argue that in order to wield policy influence, a bureaucracy requires the capacity to develop autonomous preferences (autonomy of will) and the ability to translate these preferences into action (autonomy of action) (Bauer and Ege 2014; Caughey, Cohon, and Chatfield 2009). To measure bureaucratic autonomy, we use the following eight indicators (each ranging from 0 [low] to 1 [high]), which are then combined into an unweighted additive index (ranging from 0 to 8). After the description of the individual indicators, Table 2.1 will provide a summary of the operationalization of bureaucratic autonomy.

To understand the autonomous will of IPAs, one must first consider the fact that bureaucracies are collective actors. Hence, we take into account IPAs' administrative cohesion, which depends on their staff members' ability to overcome obstacles to collective action and interact with political actors as a unified organizational entity (Mayntz 1978: 68). IPAs can be expected to be cohesive if staff members have similar national backgrounds and have been able to stay with the organization over a longer period of time. Second, the development of an autonomous will requires administrative differentiation, which allows staff members to form distinct (administrative) preferences that can potentially differ from those of the political principals. We measure this dimension by considering independent leadership (Cox 1969) and independent research capacities (Haas 1992) as two important means that facilitate the potential for administrative differentiation in IPAs. While independent leaders can be expected to defend the secretariat's position against political pressure, independent research capacities are an important means for an administration to develop (and defend) policy options that are different from those of the political actors of the IGO.

In order to be attributed *autonomous action* capacities that allow the bureaucracy to translate its (potentially distinct) preferences into action, delegation research highlights the relevance of formal powers and independent administrative resources (Hooghe and Marks 2015). The powers of IPAs culminate in the functional role of the Secretary-General (SG) as the organization's highest civil servant. These powers concern their ability to insert independent proposals into the political process and also the ability of the entire bureaucracy under the SG's leadership to sanction those who do not comply with organizational rules and norms (Joachim, Reinalda, and Verbeek 2008). Moreover, the resources of an organization need to be sufficiently high, as well as independent from its members. Staffing and funding are the most important resources of public organizations. Thus, having enough of their own staff available to work within a particular issue area and being financially independent from member states and other donors are key features in this respect (Brown 2010).

Based on these propositions, Table 2.1 summarizes the indicators used to measure autonomy. A more detailed description of the measurement is presented in Bauer and Ege (2016a, b). Combining the indicator scores into an additive index creates an autonomy continuum with two extreme poles at the end. Bureaucracies with high structural autonomy have the potential to be particularly influential during policymaking. They combine substantive executive powers and resources with a capacity for independent preference formation and internal cohesion. As such, they constitute a strong administrative counterbalance to the IGO's political sphere. Autonomous bureaucracies may use their central position to influence policymaking throughout the policy cycle, ranging from policy initiation and drafting to implementation and service delivery. Bureaucracies with low structural autonomy play a relatively passive role during policymaking and only provide technical assistance or monitor tasks either at the IGO's headquarters or in the organization's field missions and offices. This may also include executive duties that the IPA implements relatively autonomously, but only for tasks that can be clearly specified by the political principals, for example, through rule-based delegation.

### *Administrative Styles*

The concept of administrative styles emerged in the context of comparative public policy and public administration literature. Administrative styles can generally be defined as stable informal patterns that characterize the behavior and activities of public administrations in the policymaking process (Bayerlein, Knill, and Steinebach 2020; Knill 2001; Knill and Grohs 2015). Administrative styles manifest themselves in organizational routines and standardized practices and are as

Table 2.1 *Measurement of bureaucratic autonomy*

|   | Dimension | Indicator | Operationalization |
|---|---|---|---|
| Autonomy of will | Cohesion | Homogeneity of staff (nationality-based) | Ratio of ten largest nationalities (in terms of staff) to total organizational personnel |
|   |   | Administrative permanence | Ratio of staff with open-ended contracts to total number of staff<br>No mobility rules: **High**<br>Mobility is voluntary but explicitly encouraged: **Medium**<br>Mobility is mandatory: **Low** |
|   | Differentiation | Independent leadership | Share of heads of administration recruited from within the organization |
|   |   | Independent research capacities | Centrality of research bodies at different hierarchical levels<br>Existence of a research body at the department level (directly below the SG): **High**<br>Existence of two or more research bodies at the division level (two hierarchical levels below the SG): **Medium high**<br>Existence of one research body at the division level (two hierarchical levels below the SG): **Medium low**<br>No research body at division level or above: **Low** |
| Autonomy of action | Powers | Agenda competences | Degree to which the SG is involved in setting the agenda for legislative meetings<br>SG is responsible for the preparation of the draft agenda and items cannot be removed prior to the actual legislative meeting: **High**<br>SG is responsible for the preparation of the draft agenda but items can be removed prior to the actual meeting: **Medium high**<br>The executive body, not the SG, is responsible for the preparation of the draft agenda and items cannot be removed: **Medium low**<br>The executive body, not the SG, is responsible for the preparation of the draft agenda and items can be removed: **Low** |

Table 2.1 (cont.)

| Dimension | Indicator | Operationalization |
|---|---|---|
| | Sanctioning competences | *Sanctioning powers of the organization vis-à-vis its members*<br>Autonomous capacity to impose sanctions: **High**<br>Power to call for sanctions against noncompliant members: **Medium high**<br>Denial of membership benefits (e.g., voting rights and IGO services): **Medium low**<br>Only naming and shaming by issuing reports or admonitions: **Low** |
| Resources | Personnel resources | *Number of total secretarial staff per policy field*<br>Organization employs 1,500 staff or more per policy field: **High**<br>Organization employs between 1,000 and 1,499 staff per policy field: **Medium high**<br>Organization employs between 500 and 999 staff per policy field: **Medium low**<br>Organization employs less than 500 staff per policy field: **Low** |
| | Financial resources | *Degree to which the organization relies on independent sources of income*<br>Self-financing: **High**<br>Mandatory contributions: **Medium**<br>Voluntary contributions: **Low**<br>(In case an organization relies on several financial resources, we use the source with the highest share of the budget.) |

*Source:* Bauer and Ege (2017)

such distinct from the deliberate strategic behavior of an IPA's staff or bureaucratic politics (e.g., Allison 1971). Following Knill et al. (2019: 85–86, emphasis in original),

> we conceive of administrative styles as *relatively stable behavioral orientations* characterizing an organizational body. It is an institutionalized informal *modus operandi* that materializes as a guiding principle over time and by repetition, routinization, and subsequent internalization. Under conditions of uncertainty and complexity, individual bureaucrats develop routines for coping with shortages of knowledge, information-processing capacities, and time (Simon, 1997). Similarly and depending on their underlying rationale, administrators can develop and internalize behavioral patterns sought to influence their organization's policies (Knill, 2001; Wilson, 1989). We interpret these observable patterns as corresponding to an ideal typical characterization of IPAs' 'styles' in shaping IPA behavior. Rather than restricting our analytical focus on an IPA's formal position we are thus interested in the extent to which an IPA developed *informal routines* that allow it to exert influence beyond formal rules or whether its informal activities remain in line with or even behind its formal position.

The study of administrative styles originated from early attempts to "characterize and account for the significantly different ways people carry out relatively standard political/administrative tasks" (Davies 1967: 162). Under conditions of uncertainty and complexity, administrators and policymakers develop routines in order to cope with shortages of knowledge, information-processing capacities, and time (Simon 1997). At the organizational level, such coping strategies can consolidate into stable patterns of problem-solving behavior (Wilson 1989).

To measure administrative styles at the level of international organizations we analytically differentiate between different patterns of administrative involvement in the initiation, policy formulation, and implementation of policies (Bayerlein, Knill, and Steinebach 2020; Knill, Eckhard, and Grohs 2016). In each phase, we assess IPA activities along three indicators that capture both functional aspects of technically sound policymaking and political aspects that guarantee alignment with political interests of the principals from an early stage on (Aberbach, Putnam, and Rockman 1981; Mayntz and Derlien 1989). After the description of the individual indicators, Table 2.2 will provide a summary of the operationalization of administrative styles.

During the stage of policy initiation, IPAs might vary in their ambitions to come up with new policy items that should be addressed (*issue emergence*), to mobilize support for their policies (*support mobilization*), and to identify the political preferences of their principals with regard to certain initiatives (*mapping of political space*). In the drafting stage, IPAs might vary in their approach to develop policy solutions (*solution search*), their efforts placed on *internal coordination*, and the extent to which they consider the political preferences of their principals when developing their drafts (*political anticipation*). With regard to the implementation stage, we consider the extent to which IPAs make *strategic use of their formal control and sanctioning power*, their engagement in *policy evaluation*, and their *ambitions to promote IGOs' policies* in their organizational environment. Overall, we can thus identify nine activities – three for each stage of the policy cycle – in which IPAs regularly have room to maneuver. We suggest all indicators to be equally important for the assessment of an IPA's administrative style.

For each of these nine activities, we differentiate two extreme poles. One is the policy entrepreneur as stylized by Kingdon (1984) and others (Mintrom and Norman 2009), an advocate of policy proposals who shows a willingness to invest time and resources in the hope of future return. The policy entrepreneur is highly active in detecting new policy problems and bringing them to the agenda, constantly observing political opportunities, and in strategically mobilizing political or societal support to shape the political agenda. When formulating policy proposals, entrepreneurs are perfectionists in the sense that their proposals are based on a holistic triangulation of the problem at hand, the desired end, and the available resources (Mintzberg 1978). As implementers, entrepreneurs are interventionists. Although the secretariat's sanctioning powers might be limited, bureaucrats can increase their steering capacity by collecting systematic information on policy effects or by developing close relationships with involved stakeholders, interest groups, national administrations, or external experts.

On the other end of the spectrum resides the more pragmatic servant-style administration resembling Max Weber's ideal-typical conception of bureaucracy as a value-neutral machinery. Servant administrations pursue a "wait-and-see" approach by primarily responding to external policy requests instead of actively exploring windows of opportunity. When formulating policies, we can observe an instrumental and service-oriented role perception and the perpetuation of existing policies in an incremental manner (Lindblom 1959). From such a perspective, civil servants would do as requested and not question the substance of their tasks, even if they found them to be flawed. In implementation, the servant secretariat relies on a mediating approach. It refrains from observing and trying to improve compliance that goes beyond its legally specified duties, and it relies on nonhierarchical mechanisms of self-regulation. The servant-style IPAs must not necessarily be equated with suboptimal performance or the absence of intentional action per se. It is well possible that a servant-style IPA conceives of itself as a "good" and faithful servant to its political principal and acts accordingly (Boyne and Walker 2004: 240; Rainey 1997).

Knill et al. (2019) provide a more comprehensive discussion of the concept and its determinant, arguing that styles vary depending on the extent to which an

IPA is challenged externally (perceived domain challenges, perceived political challenges) and internally (lack of cognitive slack, contested belief systems). Depending on the nature of these challenges, some indicators point to a more entrepreneurial style and others point to servant behavior, thereby reflecting the overall style as being between the two extreme poles. Knill et al. (2019) further discuss how the configuration of individual indicator values can be theoretically meaningful. This means that change in styles is possible to the extent that external or internal challenges change, which should occur only gradually.

Table 2.2 summarizes the operationalization of administrative styles. Empirical data was gathered by Bayerlein, Knill, and Steinebach (2020) on the basis of semistructured expert interviews with 124 individuals at the headquarters of an IGO between 2015 and 2017 (see interview list in Table 2.A1). Each interview lasted between thirty and ninety minutes and followed the list of indicators as presented in Table 2.2. Adjustments were made depending on an interviewee's job profile. Interviews were recorded and transcribed afterward. Individual questions/statements were coded qualitatively along the operationalization in Table 2.2 on which basis each organization received one value for each indicator, ranging from low to high (see Bayerlein, Knill, and Steinebach 2020, for more details on the data and measurement). For the present purpose, we translate these measures into numerical values (low = 0, medium = 0.5, high = 1) and construct an additive index of administrative styles, with a low overall value representing a servant style and a high overall value representing an entrepreneurial style.

While there is no endogeneity problem in the measurement of the two concepts, one may find a slight conceptual overlap between the dimension "administrative differentiation" (autonomy) and "solution search" (style). Owing to the different means of data collection (staff interviews vs. structural characteristics of the IPA) this should not be much of a problem for the following comparison – also in view of the advantage of being able to systematically study how these two dimensions are related.

## 2.3 Theoretical Considerations on the Relationship between Formal and Informal Institutions

In the previous section, we suggested two systematic ways to conceive and measure formal and informal characteristics of international bureaucracies. The basis of our theoretical and analytical considerations remains, however, restricted to these concepts. In other words, no orientation emerged as to how the two spheres, the formal and the informal, relate to each other. Conceptualizing the link between the two concepts is the aim of the following paragraphs.

Table 2.2 *Measurement of administrative styles*

| Phase | Indicator | Operationalization |
|---|---|---|
| Policy initiation | *Mapping of political space* | Usually no mapping activities to investigate the IPA's principals' preferences at an early stage: ***Low*** <br> No clear pattern; occasional mapping activities to investigate the IPA's principals' preferences at an early stage: ***Medium*** <br> Usually strong mapping activities to investigate the IPA's principals' preferences at an early stage: ***High*** |
| | *Support mobilization* | Usually no mobilization activities; no active coalition-building exercises to gain external support: ***Low*** <br> No clear pattern; occasional mobilizations activities: ***Medium*** <br> Usually strong mobilizations activities; active coalition-building exercises to gain external support: ***High*** |
| | *Issue emergence* | Usually outside the bureaucracy: ***Low*** <br> No clear pattern; occasionally within the bureaucracy: ***Medium*** <br> Usually within bureaucracy: ***High*** |
| Policy drafting | *Political anticipation* | Usually no functional politicization; the IPA is routinely not sensitive to its political implications: ***Low*** <br> No clear pattern; occasional functional politicization: ***Medium*** <br> Usually strong functional politicization; the IPA is routinely very sensitive to its political implications: ***High*** |
| | *Solution search* | Usually pragmatic drafting with short-cuts or simple heuristics, settling for the first best solution: ***Low*** <br> No clear pattern; occasional systematic assessment of the underlying problems and a consideration of alternatives, settling for the optimal solution: ***Medium*** <br> Usually systematic assessment of the underlying problems and a consideration of many alternatives, settling for the optimal solution: ***High*** |

| | | |
|---|---|---|
| | Internal coordination | Usually no efforts to deviate from the default mode of negative coordination: *Low*<br>No clear pattern; Occasional efforts to deviate from the default mode of negative coordination: *Medium*<br>Usually strong effort to deviate from the default mode of negative coordination: *High* |
| Policy implementation | Strategic use of formal powers | Usually the IPA refrains from open conflicts: *Low*<br>No clear pattern; occasionally the IPA makes strategic use of its formal power: *Medium*<br>Usually the IPA makes strategic use of its formal power: *High* |
| | Policy promotion | Usually the IPA makes no efforts to strengthen the impact of organizational outputs: *Low*<br>No clear pattern; occasionally the IPA makes efforts to strengthen the impact of organizational outputs: *Medium*<br>Usually the IPA makes strong efforts to strengthen the impact of organizational outputs in every possible way: *High* |
| | Evaluation efforts | Usually the IPA barely follows the formal evaluation guidelines or does not apply them properly: *Low*<br>No clear pattern; occasionally the IPA follows the formal evaluation guidelines: *Medium*<br>Usually the IPA strongly follows the formal evaluation guidelines and makes frequent use of the institutional evaluation mechanisms: *High* |

*Source*: Bayerlein, Knill, and Steinebach (2020)

Studying the interplay between formal and informal organizational features is a long-standing and traditional research topic for organizational theorists (see, e.g., Groddeck and Wilz 2015; Tacke 2015). In the field of public administration, diverse aspects of the relationship between formal and informal features of organization have been studied, ranging from the interplay between the formal and informal accountability structures (Busuioc and Lodge 2016), and the link between formal discretion and informal behavior of street-level bureaucrats (Lipsky 1980), to the relationship of formal and actual autonomy of regulatory agencies (Jackson 2014; Maggetti 2007). In a similar vein, differentiating between formal and informal features of organizations is also prominent in IGO research (Jankauskas 2022). Martin (2006: 141), for instance, distinguishes "between formal agency, which is the amount of authority states have explicitly delegated to an I[G]O, and informal agency, which is the autonomy an I[G]O has in practice, holding the rules constant."

Yet while the distinction of formal and informal institutions can be considered as common sense in the relevant literature, the theoretical conception of the relationship between both elements is far from straightforward. In this regard, we can conceive of two scenarios that emphasize either tightly or merely loosely coupled formal and informal arrangements.

Departing from a *tight coupling* scenario, we expect that the degree of structural autonomy of an IPA should largely determine its administrative styles. In this regard, the most straightforward expectation is that higher autonomy should come along with more entrepreneurial style patterns. Yet the assumption of tight coupling of this kind would factually render the differentiation between formal and informal arrangements obsolete. If informal routines are epiphenomenal to formal institutions, there is no need to study the informal side of the story as no independent explanatory added value is to be expected. Instead, we could simply rely on structural autonomy in order to estimate the potential influence of IPAs on policymaking beyond the nation-state. To additionally look at informal routines would be superfluous.

In fact, the heavy emphasis placed on the distinction between formal and informal institutions in the literature lends strong support to assume a scenario of *loose coupling*, in which structural autonomy and administrative styles are considered as phenomena independent of each other. The justification for this view emerges from the fact that the literature emphasizes rather different factors that influence variation in terms of formal and informal arrangements. While structural autonomy, for example, is primarily explained against the background of principals' preferences, institutional path dependencies but also functionalist reasoning (Ege 2017; Hawkins et al. 2006b; Pierson 2000), informal institutions like administrative styles have their roots in factors like socialization, common professional backgrounds, and administrative perceptions, as well as narratives of

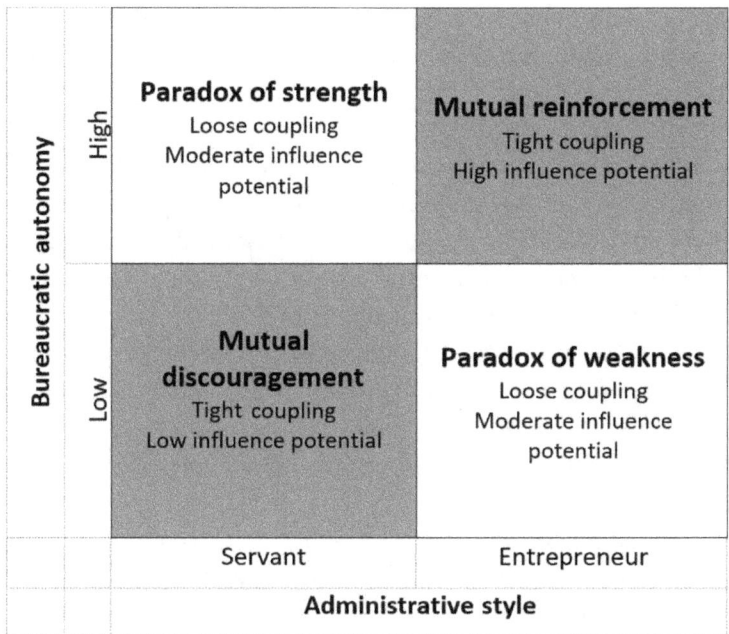

Figure 2.1 Ideal-typical configurations of formal and informal potentials of bureaucratic policy influence

external challenges through competition in organizational fields or political threat (Bayerlein, Knill, and Steinebach 2020; Knill 2001; Knill, Eckhard, and Grohs 2016). In short, the fact that different variables account for variation in formal and informal institutions should lead us to conceive of both elements as independent phenomena. Consequently, a highly autonomous IPA does not necessarily need to adopt an entrepreneurial style, while an IPA with low autonomy may not automatically adopt a servant style.

Against these considerations, we can distinguish four ideal-typical configurations of bureaucratic influence potentials. These are based on the differential relationship between formal and informal bureaucratic characteristics in the form of IPA autonomy and styles (Figure 2.1).

In line with the conceptual nature of administrative autonomy and administrative styles as outlined earlier, we expect that formal and informal arrangements can reinforce (also in terms of their mutual absence) or weaken each other with regard to an IPA's potential influence on policymaking. The highest potential for bureaucratic policy influence is expected in constellations in which high structural autonomy is paired with an entrepreneurial administrative style. By contrast, a rather low influence potential can be expected for the combination of low structural autonomy and a servant style shaping informal administrative procedures.

A moderate potential for bureaucratic influence can be expected in the two remaining constellations, which are defined by either the combination of high structural autonomy and a servant style or the combination of low structural autonomy and an entrepreneurial style.

While patterns of mutual reinforcement or discouragement at first glance seem straightforward, the other two patterns (the bottom right and the top left corner of Figure 2.1) might be characterized as rather paradoxical. As already argued elsewhere (see Knill, Eckhard, and Grohs 2016), an IPA with high structural autonomy that develops informal routines that mean the bureaucracy actually remains below its formally available influence potential reflects a *paradox of strength*. By contrast, an IPA that is formally weak but combines this with a strong entrepreneurial orientation reflects a *paradox of weakness*. We expect the potential for policy-making influence in both these cases to be moderate, given that either structural or behavioral limitations remain.

There are no reasons to assume a priori that any of these four constellations (as well as any administration between the different extreme poles) is more or less likely to emerge empirically. In particular, we should not expect the constellations *mutual reinforcement* and *mutual discouragement* to reflect more stable and more dominant constellations than any other configuration of formal and informal influence potentials. If this were the case, by contrast, we should indeed see a deterministic linkage between formal and informal arrangements – a constellation we would expect neither in light of our theory nor in view of the state of the art in IPA influence research.

A first glance at existing research findings indeed provides support for a rather unsystematic variation of formal and informal influence potentials. Without the aforementioned theoretical roadmap, one could interpret these findings as basically inconsistent. The study of national regulatory agencies is a good illustrative example here: Hanretty and Koop (2013) find support for the reinforcement hypothesis by concluding that formal statutory autonomy is an important determinant of actual independence. However, in practice, Maggetti (2007) shows that the two features are largely decoupled from each other. He concludes formal independence is neither a necessary nor a sufficient condition for explaining variations in the de facto independence of agencies.

The same can be said about research that focuses on the secretariats of IGOs. While it is argued, for instance, that "[d]ifferences in the structure of international bureaucracy afford leading officials with varying degrees of political and procedural influence over the organizations that they manage" (Manulak 2017: 6), the establishment of this link in an empirical manner remains difficult. It can thus be concluded that despite a growing body of literature on IPAs (Barnett and Finnemore 2004; Bauer, Knill, and Eckhard 2017; Biermann and Siebenhüner 2009; Johnson and Urpelainen 2014), existing research is still inconclusive regarding the extent to which, and how, the bureaucratic structure of international administrations shapes basic behavioral patterns of its staff (see Trondal 2011: 795), as well as which specific structural factors matter most for international bureaucracies' behavior and their influence on policy output (Eckhard and Ege 2016). It is the objective of the following section to investigate such configurations of informal and formal influence potentials of IPAs more systematically.

## 2.4 Empirical Assessments of the Combination of Formal and Informal Influence Potentials

Based on the operationalization of autonomy and styles as outlined above, we have gathered and published empirical data on a large range of IPAs (Bauer and Ege 2016a; Bayerlein, Knill, and Steinebach 2020; Enkler et al. 2017; Knill, Eckhard, and Grohs 2016). Our empirical data on structural autonomy and administrative styles spans nine IPAs: ILO, UNESCO, OECD, OSCE, WHO, FAO, IOM, UNHCR, and IMF. While none of these IPAs are purely environmental bureaucracies, three have at least some responsibilities in environmental issues. This is the case with the FAO, which is, for instance, involved in the multistakeholder initiative "Partnership on the environmental benchmarking of livestock supply chains" (LEAP) that aims to improve the environmental performance of livestock supply chains.[1] Via its natural science sector, UNESCO is also active in environmental issues covering water, ecological science, and earth science.[2] Finally, the OECD collects a variety of data on environmental issues in its member states and offers it expertise on topics ranging from climate change to biodiversity.[3]

While we do not claim that this sample is representative in a general sense, it includes IPAs with diverse values in many of the dimensions that are usually highlighted as theoretically important – such as membership in the UN system, budget size, number of staff, headquarter or field presence, and policy fields. In the context of this book's environmental focus, this case selection allows us to compare environmentally active administrations with other IPAs in order to find out if they are characterized by common empirical configurations of style and autonomy.

Figure 2.2 summarizes our aggregate autonomy and style scores for the nine IPAs. Based on their values, we can establish to which of the four theoretical clusters an IPA belongs. While there is no mathematically exact way of doing this,

---

[1] www.fao.org/partnerships/leap/en/  [2] www.unesco.org/new/en/natural-sciences/environment/
[3] www.oecd.org/environment/

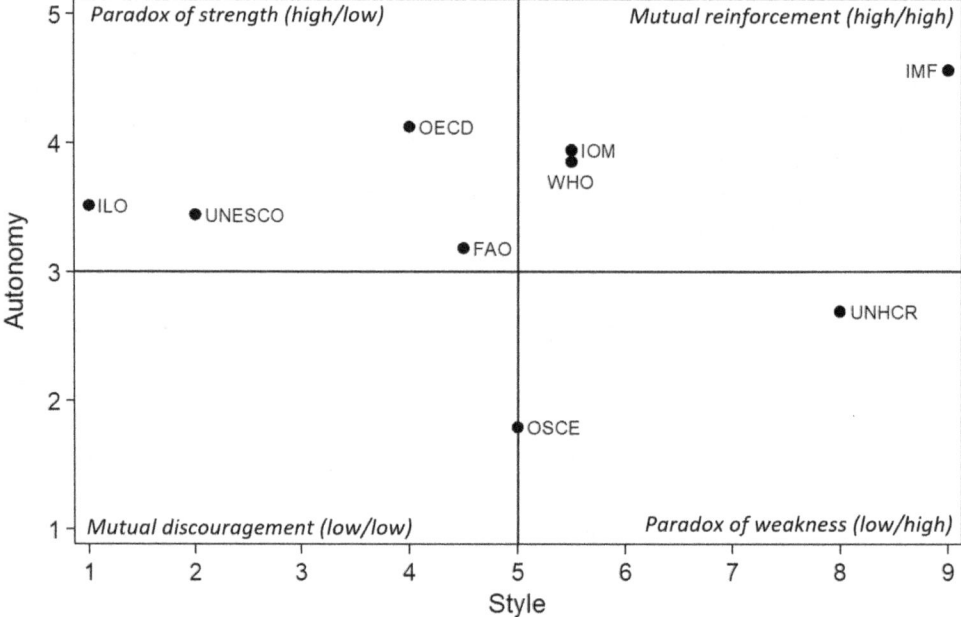

Figure 2.2 Empirical configurations of formal and informal potentials of bureaucratic policy influence

we see on the basis of their distance to one another that IPAs are represented in all fields except the area of mutual discouragement (low/low). We now highlight some exemplary quotes drawn from our interviews to illustrate how – in the context of varying degrees of autonomy – such strategic entrepreneurial or servant behavior plays out empirically in each of the quadrants.

***Paradox of weakness – combining low autonomy with an entrepreneurial style***: The UNHCR administration's autonomy is restricted in many ways (visible, e.g., in its low staff homogeneity and low administrative permanence, a lack of centralized research capacities, and weak sanctioning capacities), which is also reflected in the way bureaucrats describe their relation to member states. One interviewee, for example, said: "It would be, in my view, pretty unlikely that we would develop a policy without sufficient consultations either internally or externally. ... After all, we are dependent on the financing of some twenty countries around the world. You can't ignore your stakeholders" (UNHCR 13). Despite their financial dependence and limited structural autonomy, UNHCR staff coherently emphasize that they are not afraid of taking a clear policy position that at times even clashes with key donor interests: "We are not at all averse to conflict. Our first orientation is towards our mandate. That is the role that has been given to us and that we need to fulfil" (UNHCR 10). The key to UNHCR's entrepreneurial spirit, and this tells much about how IPA policy influence plays out in practice, is a focus on informal bargaining: "Much of our work goes on behind the scenes. When UNHCR makes a public statement of criticism of a government, it's because we have exhausted each and every level before arriving at that point" (UNHCR 1). This implies that despite their relative restrictive formal powers, UNHCR's IPA has developed a track record of good bilateral relations and informal partnership with many states that allows them to informally influence policy "behind the scenes."

*Paradox of strength – combining high autonomy with a servant style:* The ILO, UNESCO, OECD, and with limitations also the FAO are examples for the opposite scenario of loose coupling with relatively substantial autonomy and servant-style behavioral patterns. For example, the ILO's autonomy relies mainly on its independent leadership, centralized research divisions, and substantial personnel and financial resources. Because of their comparatively autonomous status, ILO bureaucrats perceive themselves as relatively unchallenged. As one staff member said with a reference to member states: "They need you. They cannot decide not to work with you. ... They can't afford to do that alone. They need the ILO, they need the expertise" (ILO 5). This one-sided dependency allows the bureaucracy to take a back seat instead of actively promoting their own agenda: "If [you ask] most of my colleagues 'how do we sell ourselves?' they won't know. Like they have no reason or objective to sell ILO to anybody" (ILO 14). The situation is similar within UNESCO, as one interviewee who explained UNESCO's policy planning process detailed: "[T]he secretariat is involved but not in terms of the design process. I don't think it is our role.... It is a country-driven process" (UNESCO 3). It is similar in the OECD too, where an official said that it "is very important for us to keep regular contact with the member states ... you can really see what the problems and topics are and then we make our proposals for the work program out of that" (OECD 7). Bureaucrats in all four organizations thus wait for request instead of developing their own ideas and convincing others to turn them into policy or to implement them. One interviewee said with an eye on their policy engagement that "[it is] less mapping of the political space. It is mostly responding to requests" (ILO 4). Another respondent said that "we are the pen holders, we do as told" (ILO 10). Finally, the servant strategy implies that all IPAs refrain from exerting pressure on member states or taking sides. For instance, when it comes to the implementation of their recommendations, the OECD remains very soft: "[A]t the moment we sort of hand over the report and we don't come back to it" (OECD 8). ILO bureaucrats avoid taking sides by fostering one or the other policy position or brokering coalitions behind the scenes: "I wouldn't say we are trying to build up pressure. In fact, we are often seen as a neutral party in these kinds of things. That's the added value of the ILO … we are not promoting any particular agenda" (ILO 14). A UNESCO staff member said that it is often "difficult to change partners so you … simply report that it unfortunately wasn't possible" (UNESCO 5). All in all, this shows that these IPAs do not exploit the potential influence they gain through their structural autonomy.

***Mutual reinforcement – combining high autonomy with an entrepreneurial style***: The IMF, which is based in Washington, DC, is the clearest example in our sample of an IPA with both an entrepreneurial style and high autonomy values. The IMF is responsible for overseeing and safeguarding the stability of the international monetary and financial system and has 189 member countries (International Monetary Fund 2017). The IMF administration's autonomy is, in comparison to others, mainly a result of high administrative permanence, strong research capacities, relatively strong sanctioning capacities, and independent financing. The IMF has significant formal powers (Lang and Presbitero 2018), which is also how its personnel perceive it: "While [other IGOs] may actually be much stronger on some topics than the Fund, for example climate change, the Fund still gets more attention and recognition in this area.… This really shows the power of the Fund" (IMF 7). Interestingly, this does not coincide with a behavioral pattern of neutrality and response, as in the ILO or UNESCO. Instead, the IMF IPA is characterized by an entrepreneurial style with respondents frankly admitting that they do take sides and promote certain policy positions: "[T]he only way to implement (certain policies) is to convince the authorities that it is something good for them to do. We also try to build a consensus around certain policies, and if most countries are on board with it we can tell the remaining ones 'you are the only ones not doing this'" (IMF 2). Another respondent said that

at the IMF it is very different from my previous work [at another IGO]. There, we may not have liked the decisions the principals made, but we knew that was the place where the decisions were made. The way I see it here, staff think they could make the decisions themselves, so why should they trust the top management or the Board [i.e., member states, the authors] to make the right decisions?

*(IMF 3)*

***Mutual discouragement – combining low autonomy with a servant style:*** Even though the OSCE with its low autonomy and medium entrepreneurism comes closest to this configuration, a clear empirical manifestation of this ideal-type is missing in our sample. It is puzzling that the configuration that most closely resembles the idea of IGOs that has for decades been predominant in theorizing in international relations is absent empirically. We can only speculate as to whether this is a peculiarity of our case selection or indicative of a broader phenomenon. While this result may raise doubts about the representativeness of our sample, it is also possible that an IPA in this quadrant is generally of limited use for IGO members – as a certain degree of autonomy is a functional requirement for an IPA to fulfil its tasks in the first place (Hawkins et al. 2006b: 13; Mayntz 1978: 66–67). Thus, in the absence of structural autonomy, an IPA may be able to actively compensate this precarious situation by developing a particularly entrepreneurial behavior to eventually justify its existence to the members. Thus, the lack of empirical cases in this field could indicate that informal and formal factors of IPAs are not fully independent of each other but that the informal administrative style can be interpreted as strategic reaction to predetermined formal context factors. This would explain why we find this "paradoxical" combination. To further substantiate this argument, however, the sample needs to be extended to include more low-autonomy organizations.

## 2.5 Conclusion

In this chapter we argued that if a theory of international bureaucratic influence is the aim, we have to disentangle the relationship between formal and informal administrative characteristics in view of the resulting potential for administrative impact on policymaking. Therefore, we revisited the controversial debate about the relationship between formal and informal features of public administrations. We presented concepts to identify and measure structural autonomy (an example of formal characteristics) and administrative styles (an example of informal routines) of international public administrations. By mapping empirical intensities of structural autonomy as well as the occurrence of (entrepreneurial or servant-like) styles, we identified four constellations of the relationship between these formal and informal characteristics. More specifically, we have asked whether the relationship is characterized predominantly by tight or loose coupling of the two concepts and applied our theoretical considerations by means of an empirical assessment of autonomy and styles in nine IPAs. Moreover, we used interview quotations from the UNHCR, ILO, and IMF administrations to illustrate the empirical existence of these combinations.

Our findings display no dominant pattern in which formal autonomy and informal styles are linked. In the majority of our cases, however, we observe a loose coupling of the two features, implying what we call the paradox of strength and weakness rather than mutual reinforcement or discouragement. Thus, our findings indicate that formal autonomy does not determine administrative styles. Rather formal autonomy and administrative styles are best considered as influence potentials that evolve and operate independent of each other. Consequently, we cannot simply rely on structural autonomy in order to estimate the potential influence of IPAs. Instead, in order to explain IPA influence on policymaking beyond the nation-state, the two aspects need to be conceptualized separately and linked empirically. These findings have important implications. First, the case of

UNHCR indicates that IPAs are capable of entrepreneurial informal reactions to situations of precarious structural autonomy. This supports previous arguments made with regard to the OSCE (Knill, Eckhard, and Grohs 2016) and provides further evidence that such a paradox of weakness may indeed be a more common feature of structurally weak administrations. Second, our insights have important implications for the design of accountability mechanisms in view of optimizing bureaucratic control in the international sphere. The finding that formal autonomy and informal styles work relatively independent of each other emphasizes that formal control and oversight in IGOs (Grigorescu 2010) may be effective only if supplemented by more informal means of securing accountability. Otherwise, member states as the collective principals may indeed be faced with a runaway bureaucracy (Elsig 2007). A growing body of research on the ways member states seek representation in IPA staff bodies and thereby enact control and influence (Eckhard and Dijkstra 2017; Eckhard and Steinebach 2021; Manulak 2017; Urpelainen 2012) is indicative of this argument. Third, and this is particularly relevant in the context of this book, our findings show that all three environmentally active IPAs studied here are characterized by a loose coupling of style and autonomy, leading to what we describe as a paradox of strength. While such a combination of high autonomy and a servant style may result in a formally influential but often rather passive role played by the three IPAs, this finding suggests that it is particularly important for other organizations populating the global administrative space in environmental governance to step in and take initiative. The chapters of this volume show that multilateral environmental convention secretariats seem to have especially taken on this challenge and over time have become more entrepreneurial, attention-seeking, and influential.

Even though we did not study these secretariats in this chapter, we argue that owing to the nature of (global) environmental policy, investigating environmental convention secretariats' autonomy and styles (as well as the relationship between the two means of influence) is promising. First efforts to apply the concepts presented here have been made. For instance, a recent study analyzes the administrative styles of the climate secretariat in the run-up to the Paris Agreement and during the course of its implementation (Saerbeck et al. 2020).

In this chapter, we focused on intrabureaucratic factors and the question of how they are related to each other in view of the administrative potential to influence international policymaking. Thus, a word of caution is in order. Studying bureaucratic influence by putting formal and informal organizational features center stage is not to deny that there are many other factors such as organizational leadership, member states' political agendas, situational staff preferences, the structure of the underlying problem, or external events that need to be considered, if in empirical cases the concrete or de facto influence of an international bureaucracy is to be established in particular cases. In their seminal work on the influence of the managers of global change, Biermann and Siebenhüner (2009) have provided a realistic design for studying such bureaucratic impact covering a broad range of factors. Even more than a decade later, empirical research can rely on their conceptual blueprint. Our modest contribution in this chapter is intended to complement this work by advancing on the intraorganizational side of the story. Thus, future research may want to investigate why and when reinforcement and discouragement takes place and how the different constellations can be interpreted in terms of bureaucratic influence. While we could only hypothesize how the two features impact on bureaucratic influence, this nexus should be explored empirically by conceptualizing influence as a separate dependent variable.

## Annex

Table 2.A1 *Interview list for the measurement of administrative styles as presented in Figure 2.2*

| IGO No. Interviewee's position Date | IGO No. Interviewee's position Date |
|---|---|
| IMF 1 Economist February 2015 | IOM 1 Associate Expert, MECC May 2015 |
| IMF 2 Deputy Division Head December 2015 | IOM 2 Disaster Risk Reduction, livelihoods and urbanization expert May 2015 |
| IMF 3 Senior Economist December 2015 | IOM 3 Head of International Processes May 2015 |
| IMF 4 Economist December 2015 | IOM 4 Migration Policy Officer, Global Processes Unit May 2015 |
| IMF 5 Division Chief December 2015 | IOM 5 Consultant, Migrants in Countries in Crisis Initiative May 2015 |
| IMF 6 Senior Economist December 2015 | IOM 6 Global Project Coordinator, Migration, Environment & Climate Change May 2015 |
| IMF 7 Advisor December 2015 | IOM 7 Head, Immigration and Border Management May 2015 |
| IMF 8 Executive Director December 2015 | IOM 8 Chief of Staff June 2016 |
| IMF 9 Economist December 2015 | IOM 9 Director of International Cooperation and Processes June 2016 |
| IMF 10 Economist December 2015 | IOM 10 Senior Labour Migration Specialist June 2016 |
| IMF 11 Economist December 2015 | IOM 11 Global CCCM Cluster Coordinator June 2016 |
| IMF 12 Assistant Director December 2015 | IOM 12 Research Assistant June 2016 |
| IMF 13 Assistant Director December 2015 | IOM 13 Special Policy Advisor to the Director General June 2016 |
| IMF 14 Division Chief December 2015 | |
| IMF 15 Advisor December 2015 | |
| ILO 1 Senior Advisor May 2015 | |

Table 2.A1 (cont.)

| IGO No. Interviewee's position Date | IGO No. Interviewee's position Date |
|---|---|
| ILO 2 Programme and Operations Officer May 2015 | IOM 14 Global CCCM Civil Protection Specialist June 2016 |
| ILO 3 Programme and Operations Officer May 2015 | IOM 15 Labour Mobility & Human Development, Head of Division June 2016 |
| ILO 4 Senior Advisor May 2015 | IOM 16 Chief of Mission Libya June 2016 |
| ILO 5 Technical Officer May 2015 | OECD 1 Senior Project Manager February 2015 |
| ILO 6 Senior Advisor May 2015 | |
| ILO 7 Programme and Operations Officer June 2016 | OECD 2 Economist February 2015 |
| | OECD 3 Senior Analyst February 2015 |
| ILO 8 Programme and Operations Officer June 2016 | OECD 4 Director February 2015 |
| | OECD 5 Senior Analyst February 2015 |
| ILO 9 Technical Officer June 2016 | OECD 6 Senior Economist February 2015 |
| ILO 10 Country Director June 2016 | OECD 7 Principal Administrator February 2015 |
| ILO 11 Technical Officer June 2016 | |
| ILO 12 Technical Officer June 2016 | OECD 8 principal administrator February 2015 |
| ILO 13 Senior Advisor June 2016 | OECD 9 Senior Economist February 2015 |
| ILO 14 Programme and Operations Officer June 2016 | OECD 10 Policy Analyst February 2015 |
| | OECD 11 Senior Economist February 2015 |
| FAO 1 Technical Officer May 2015 | OECD 12 Senior economist, head of unit February 2015 |
| FAO 2 Consultant June 2015 | |
| FAO 3 Programme Officer June 2015 | OECD 13 Senior economist, head of unit February 2015 |
| FAO 4 Policy Advisor June 2015 | |
| FAO 5 Consultant May 2015 | OECD 14 Senior Economist February 2015 |
| FAO 6 Programme Officer May 2015 | OECD 15 Policy Analyst February 2015 |
| FAO 7 Communication Officer May 2015 | OECD 16 Deputy Director February 2015 |
| | OECD 17 Head February 2015 |
| FAO 8 Senior Official June 2015 | OECD 18 Senior Analyst Financial and Enterprise Affairs February 2015 |
| FAO 9 Professional Officer June 2015 | |
| FAO 10 Programme Officer June 2015 | OECD 19 Senior Analyst Migration February 2015 |
| FAO 11 Consultant May 2015 | |
| FAO 12 Technical Officer June 2015 | OECD 20 Senior Analyst February 2015 |
| FAO 13 Technical Officer June 2015 | OECD 21 Senior Analyst February 2015 |
| FAO 14 Consultant June 2015 | OECD 22 Policy Analyst February 2015 |
| FAO 15 Programme Officer May 2015 | OECD 23 Policy Analyst February 2015 |
| FAO 16 Consultant June 2015 | UNESCO 1 Director February 2015 |
| FAO 17 Consultant May 2015 | UNESCO 2 Programme Officer February 2015 |
| FAO 18 Professional Officer May 2015 | UNESCO 3 Programme Officer February 2015 |
| FAO 19 Senior Official June 2015 | UNESCO 4 Programme Officer February 2015 |
| OSCE 1 Transnational Threats Department March 2017 | UNESCO 5 Assistant Programme Specialist February 2015 |
| OSCE 2 Office of the Co-ordinator of OSCE Economic and Environmental Activities March 2017 | UNESCO 6 Programme Officer February 2015 |
| | UNESCO 7 Programme Officer February 2015 |
| | UNESCO 8 Deputy Ambassador February 2015 |
| OSCE 3 Conflict Prevention Centre March 2017 | |
| | UNESCO 9 Chief of Programme April 2016 |
| OSCE 4 Conflict Prevention Centre March 2017 | UNESCO 10 Chief of Section April 2016 |
| | UNESCO 11 Director April 2016 |

Table 2.A1 (cont.)

| IGO No. Interviewee's position Date | IGO No. Interviewee's position Date |
|---|---|
| OSCE 5 Conflict Prevention Centre March 2017 | UNESCO 12 Senior Analyst April 2016 |
| OSCE 6 Conflict Prevention Centre March 2017 | UNHCR 1 Senior Official May 2016 |
| | UNHCR 10 Senior Official May 2016 |
| OSCE 7 OSCE Secretariat Staff Committee March 2017 | UNHCR 13 Head of Unit March 2017 |
| | WHO 1 Embassy expert June 2016 |
| OSCE 8 Office of Internal Oversight March 2017 | WHO 2 Director June 2016 |
| | WHO 3 Team Lead June 2016 |
| OSCE 9 Transnational Threats Department March 2017 | WHO 4 Team Lead June 2016 |
| | WHO 5 Director June 2016 |
| OSCE 10 Transnational Threats Department March 2017 | WHO 6 Technical Officer June 2016 |
| | WHO 7 Director June 2016 |
| OSCE 11 Department of Human Resources March 2017 | WHO 8 Coordinator June 2016 |
| | WHO 9 Director July 2016 |
| OSCE 12 Department of Human Resources March 2017 | WHO 10 Director July 2016 |
| OSCE 13 Department of Human Resources March 2017 | |
| OSCE 14 Department of Human Resources March 2017 | |
| OSCE 15 Office of the Secretary-General March 2017 | |

## References

Abbott, K. W. and Snidal, D. (1998). Why States Act through Formal International Organizations, *Journal of Conflict Resolution 42* (1): 3–32.

Aberbach, J. D., Putnam, R. D., and Rockman, B. A. (1981). *Bureaucrats and Politicians in Western Democracies*, Cambridge, MA: Harvard University Press.

Allison, G. T. (1971). *Essence of Decision – Explaining the Cuban Missile Crisis*, Boston, MA: Little, Brown and Co.

Barnett, M. and Finnemore, M. (2004). *Rules for the World: International Organizations in Global Politics*, Ithaca, NY: Cornell University Press.

Bauer, M. W. and Ege, J. (2014). The Autonomy of International Bureaucracies. In S. Kim, S. Ashley, and W. H. Lambright (eds.), *Public Administration in the Context of Global Governance*, Cheltenham: Edward Elgar, 63–84.

Bauer, M. W. and Ege, J. (2016a). Bureaucratic Autonomy of International Organizations' Secretariats, *Journal of European Public Policy 23* (7): 1019–1037. DOI: 10.1080/13501763.2016.1162833.

Bauer, M. W. and Ege, J. (2016b). *Conceptualizing and Measuring the Bureaucratic Autonomy of International Public Administrations* (Working Paper No. 8), Speyer.

Bauer, M. W. and Ege, J. (2017). A Matter of Will and Action: The Bureaucratic Autonomy of International Public Administrations. In M. W. Bauer, C. Knill, and S. Eckhard (eds.), *International Bureaucracy: Challenges and Lessons for Public Administration Research*, Basingstoke: Palgrave Macmillan, 13–41.

Bauer, M. W., da Conceição-Heldt, E., and Ege, J. (2015). Autonomiekonzeptionen internationaler Organisationen im Vergleich, *Politische Vierteljahresschrift* (Sonderheft 49), Baden Baden: Nomos, 28–53.

Bauer, M. W., Knill, C., and Eckhard, S. (eds.) (2017). *International Bureaucracy: Challenges and Lessons for Public Administration Research*, Basingstoke: Palgrave Macmillan.

Bauer, M. W., Eckhard, S., Ege, J., and Knill, C. (2017). A Public Administration Perspective on International Organizations. In M. W. Bauer, C. Knill, and S. Eckhard (eds.), *International Bureaucracy: Challenges and Lessons for Public Administration Research*, Basingstoke: Palgrave Macmillan, 1–12.

Bayerlein, L., Knill, C., and Steinebach, Y. (2020). *A Matter of Style? Organizational Agency in Global Public Policy*, Cambridge Studies in Comparative Public Policy, Cambridge: Cambridge University Press.

Biermann, F. and Siebenhüner, B. (eds.) (2009). *Managers of Global Change: The Influence of International Environmental Bureaucracies*. Cambridge, MA: MIT Press.

Biermann, F., Siebenhüner, B., Bauer, S., et al. (2009). Studying the Influence of International Bureaucracies: A Conceptual Framework. In F. Biermann and B. Siebenhüner (eds.), *Managers of Global Change: The Influence of International Environmental Bureaucracies,* Cambridge, MA: MIT Press, 37–74.

Boyne, G. A. and Walker, R. M. (2004). Strategy Content and Public Service Organizations, *Journal of Public Administration Research and Theory* 14 (2): 231–252. DOI: 10.1093/jopart/muh015.

Brown, R. L. (2010). Measuring Delegation, *The Review of International Organizations* 5 (2): 141–175. DOI: 10.1007/s11558-009-9076-3.

Busuioc, E. M. and Lodge, M. (2016). The Reputational Basis of Public Accountability, *Governance* 29 (2): 247–263. DOI: 10.1111/gove.12161.

Caughey, D., Cohon, A., and Chatfield, S. (2009). Defining, Measuring, and Modeling Bureaucratic Autonomy. *Paper presented at the Midwest Political Science Association Annual Meeting, Chicago, IL.*

Cox, R. W. (1969). The Executive Head: An Essay on Leadership in International Organization, *International Organization* 23 (2): 205–230.

Davies, A. F. (1967). The Concept of Administrative Styles, *Australian Journal of Public Administration* 26 (2): 162–173.

Eckhard, S. and Dijkstra, H. (2017). Contested Implementation: The Unilateral Influence of Member States on Peacebuilding Policy in Kosovo, *Global Policy* 8: 102–112. DOI: 10.1111/1758-5899.12455.

Eckhard, S. and Ege, J. (2016). International Bureaucracies and Their Influence on Policy-Making: A Review of Empirical Evidence, *Journal of European Public Policy* 23 (7): 960–978. DOI: 10.1080/13501763.2016.1162837.

Eckhard, S. and Steinebach, Y. (2021). Staff Recruitment and Geographical Representation in International Organizations. *International Review of Administrative Sciences* 87(4): 701–717. doi:10.1177/00208523211031379.

Ege, J. (2017). Comparing the Autonomy of International Public Administrations: An Ideal-Type Approach, *Public Administration* 95 (3): 555–570. DOI: 10.1111/padm.12326.

Ege, J. and Bauer, M. W. (2013). International Bureaucracies from a Public Administration and International Relations Perspective. In B. Reinalda (ed.), *Routledge Handbook of International Organization*, London: Routledge, 135–148.

Ege, J., Bauer, M. W., and Wagner, N. (2020). Improving Generalizability in Transnational Bureaucratic Influence Research: A (Modest) Proposal, *International Studies Review* 22 (3): 551–575. DOI: 10.1093/isr/viz026.

Ege, J., Bauer, M. W., and Wagner, N. (2021). How Do International Bureaucrats Affect Policy Outputs? Studying Administrative Influence Strategies in International Organizations, *International Review of Administrative Sciences* 87 (4): 737–754. DOI: 10.1177/00208523211000109.

Ege, J., Bauer, M. W., Wagner, N., and Thomann, E. (2023). Under What Conditions Does Bureaucracy Matter in the Making of Global Public Policies?, *Governance* 36 (4): 1313–1333. DOI: 10.1111/gove.12741.

Elsig, M. (2007). *The World Trade Organization's Bureaucrats: Runaway Agents or Masters Servants?* Swiss National Center for Competence in Research. Working Paper No. 2007/19.

Enkler, J., Schmidt, S., Eckhard, S., Knill, C., and Grohs, S. (2017). Administrative Styles in the OECD: Bureaucratic Policy-Making beyond Formal Rules, *International Journal of Public Administration* 40 (8): 637–648. DOI: 10.1080/01900692.2016.1186176.

Grigorescu, A. (2010). The Spread of Bureaucratic Oversight Mechanisms across Intergovernmental Organizations, *International Studies Quarterly* 54 (3): 871–886. DOI: 10.1111/j.1468-2478.2010.00613.x.

Groddeck, V. v. and Wilz, S. M. (eds.) (2015). *Organisationssoziologie. Formalität und Informalität in Organisationen*, Wiesbaden: Springer VS.

Haas, P. M. (1992). Introduction: Epistemic Communities and International Policy Coordination, *International Organization* 46 (1): 1–35.

Haftel, Y. Z. and Thompson, A. (2006). The Independence of International Organizations: Concept and Applications, *Journal of Conflict Resolution* 50 (2): 253–275.

Hanretty, C. and Koop, C. (2013). Shall the Law Set Them Free? The Formal and Actual Independence of Regulatory Agencies, *Regulation and Governance* 7 (2): 195–214. DOI: 10.1111/j.1748-5991.2012.01156.x.

Hawkins, D. G., Lake, D. A., Nielson, D. L., and Tierney, M. J. (eds.) (2006a). *Delegation and Agency in International Organizations*, Cambridge, MA: Cambridge University Press.

Hawkins, D. G., Lake, D. A., Nielson, D. L., and Tierney, M. J. (2006b). Delegation under Anarchy: States, International Organizations, and Principal-Agent Theory. In D. G. Hawkins, D. A. Lake, D. L. Nielson, and M. J. Tierney (eds.), *Delegation and Agency in International Organizations*, Cambridge, MA: Cambridge University Press, 3–38.

Hooghe, L. and Marks, G. (2015). Delegation and Pooling in International Organizations, *The Review of International Organizations* 10 (3): 305–328.

Hooghe, L., Marks, G., Lenz, T., et al. (2017). *Measuring International Authority: A Postfunctionalist Theory of Governance, Volume III*, Oxford: Oxford University Press.

International Monetary Fund (2017). About the IMF. www.imf.org/external/about.htm

Jackson, C. (2014). Structural and Behavioural Independence: Mapping the Meaning of Agency Independence at the Field Level, *International Review of Administrative Sciences* 80 (2): 257–275. DOI: 10.1177/0020852313514525.

Jankauskas, V. (2022). Delegation and Stewardship in International Organizations, *Journal of European Public Policy* 29 (4): 568–588. DOI: 10.1080/13501763.2021.1883721.

Joachim, J., Reinalda, B., and Verbeek, B. (eds.) (2008). *International Organizations and Implementation: Enforcers, Managers, Authorities?*, London: Routledge.

Johnson, T. (2013). Institutional Design and Bureaucrats' Impact on Political Control, *The Journal of Politics* 75 (1): 183–197. DOI: 10.1017/S0022381612000953.

Johnson, T. and Urpelainen, J. (2014). International Bureaucrats and the Formation of Intergovernmental Organizations: Institutional Design Discretion Sweetens the Pot, *International Organization* 68 (1): 177–209. DOI: 10.1017/S0020818313000349.

Kingdon, J. W. (1984). *Agendas, Alternatives, and Public Policies*, Boston, MA: Little, Brown and Co.

Knill, C. (2001). *The Europeanisation of National Administrations: Patterns of Institutional Change and Persistence*, Themes in European Governance, Cambridge, NY: Cambridge University Press.

Knill, C. and Bauer, M. W. (2016). Policy-Making by International Public Administrations – Concepts, Causes and Consequences: Introduction to the Special Issue: Governance by International Public Administrations? Tools of Bureaucratic Influence and Effects on Global Public Policies, *Journal of European Public Policy 23* (7): 949–959. DOI: 10.1080/13501763.2016.1168979.

Knill, C. and Grohs, S. (2015). Administrative Styles of EU Institutions. In M. W. Bauer and J. Trondal (eds.), *The Palgrave Handbook on the European Administrative System*, Basingstoke: Palgrave Macmillan, 93–107.

Knill, C., Eckhard, S., and Grohs, S. (2016). Administrative Styles in the European Commission and the OSCE Secretariat: Striking Similarities despite Different Organizational Settings, *Journal of European Public Policy 23* (7): 1057–1076. DOI: 10.1080/13501763.2016.1162832.

Knill, C., Bayerlein, L., Enkler, J., and Grohs, S. (2019). Bureaucratic Influence and Administrative Styles in International Organizations, *The Review of International Organizations* 14 (1): 83–106. DOI: 10.1007/s11558-018-9303-x.

Koremenos, B., Lipson, C., and Snidal, D. (2001). The Rational Design of International Institutions, *International Organization 55* (4): 761–799. DOI: 10.1162/002081801317193592.

Lall, R. (2017). Beyond Institutional Design: Explaining the Performance of International Organizations, *International Organization 71* (2): 245–280. DOI: 10.1017/S0020818317000066.

Lang, V. F. and Presbitero, A. F. (2018). Room for Discretion? Biased Decision-Making in International Financial Institutions, *Journal of Development Economics 130*: 1–16. DOI: 10.1016/j.jdeveco.2017.09.001.

Liese, A. and Weinlich, S. (2006). Die Rolle von Verwaltungsstäben in internationalen Organisationen, Lücken, Tücken und Konturen eines (neuen) Forschungsgebiets. *Politische Vierteljahresschrift 37*: 491–524.

Lindblom, C. E. (1959). The Science of "Muddling Through," *Public Administration Review 19* (2): 79–88. DOI: 10.2307/973677.

Lipsky, M. (1980). *Street-Level Bureaucracy: Dilemmas of the Individual in Public Services*, New York: Russell Sage Foundation. www.worldcat.org/oclc/6209977

Maggetti, M. (2007). De Facto Independence after Delegation: A Fuzzy-Set Analysis, *Regulation & Governance 1* (4): 271–294. DOI: 10.1111/j.1748-5991.2007.00023.x.

Maggetti, M. and Verhoest, K. (2014). Unexplored Aspects of Bureaucratic Autonomy: A State of the Field and Ways Forward, *International Review of Administrative Sciences 80* (2): 239–256. DOI: 10.1177/0020852314524680.

Manulak, M. W. (2017). Leading by Design: Informal Influence and International Secretariats, *The Review of International Organizations 12* (4): 497–522. DOI: 10.1007/s11558-016-9245-0.

Martin, L. L. (2006). Distribution, Information, and Delegation to International Organizations: The Case of IMF Conditionality. In D. G. Hawkins, D. A. Lake, D. L. Nielson, and M. J. Tierney (eds.), *Delegation and Agency in International Organizations*, Cambridge, MA: Cambridge University Press, 140–164.

Mayntz, R. (1978). *Soziologie der öffentlichen Verwaltung*, Heidelberg: C. F. Müller.

Mayntz, R. and Derlien, H.-U. (1989). Party Patronage and Politicization of the West German Administrative Elite 1970–1987: Toward Hybridization?, *Governance 2* (4): 384–404. DOI: 10.1111/j.1468-0491.1989.tb00099.x.

McCubbins, M. D., Noll, R. G., and Weingast, B. R. (1989). Structure and Process, Politics and Policy: Administrative Arrangements and the Political Control of Agencies, *Virginia Law Review 75* (2): 431–482. DOI: 10.2307/1073179.

Mintrom, M. and Norman, P. (2009). Policy Entrepreneurship and Policy Change, *Policy Studies Journal 37* (4): 649–667.

Mintzberg, H. (1978). Patterns in Strategy Formation, *Management Science 24* (9): 934–948. DOI: 10.1287/mnsc.24.9.934.

Nay, O. (2012). How Do Policy Ideas Spread among International Administrations? Policy Entrepreneurs and Bureaucratic Influence in the UN Response to AIDS, *Journal of Public Policy 32* (1): 53–76. DOI: 10.1017/S0143814X11000183.

Oksamytna, K. (2018). Policy Entrepreneurship by International Bureaucracies: The Evolution of Public Information in UN Peacekeeping, *International Peacekeeping 25* (1): 79–104. DOI: 10.1080/13533312.2017.1395286.

Pierson, P. (2000). Increasing Returns, Path Dependence, and the Study of Politics, *American Political Science Review 94* (2): 251–268.

Rainey, H. G. (1997). *Understanding and Managing Public Organizations*, 2nd ed., The Jossey-Bass Nonprofit & Public Management Series, San Francisco: Jossey-Bass.

Reinalda, B. and Verbeek, B. (2004). The Issue of Decision Making within International Organizations. In B. Reinalda and B. Verbeek (eds.), *Decision Making within International Organizations*, London: Routledge, 9–41.

Saerbeck, B., Well, M., Jörgens, H., Goritz, A., and Kolleck, N. (2020). Brokering Climate Action: The UNFCCC Secretariat between Parties and Nonparty Stakeholders, *Global Environmental Politics 20* (2): 105–127. DOI: 10.1162/glep_a_00556.

Simon, H. A. (1997). *Administrative Behavior: A Study of Decision-Making Processes in Administrative Organizations*, 4th ed., New York: Free Press. www.worldcat.org/oclc/35159347

Skovgaard, J. (2017). The Devil Lies in the Definition: Competing Approaches to Fossil Fuel Subsidies at the IMF and the OECD, *International Environmental Agreements: Politics, Law and Economics 17* (3): 341–353. DOI: 10.1007/s10784-017-9355-z.

Stone, D. and Ladi, S. (2015). Global Policy and Transnational Administration, *Public Administration 4*: 839–855.

Stone, D. and Moloney, K. (eds.) (2019). *Oxford Handbooks: The Oxford Handbook of Global Policy and Transnational Administration*, 1st ed., Oxford: Oxford University Press.

Stone, R. W. (2011). *Controlling Institutions: International Organizations and the Global Economy*, Cambridge: Cambridge University Press.

Tacke, V. (2015). Formalität und Informalität – Zu einer klassischen Unterscheidung der Organisationssoziologie. In V. v. Groddeck and S. M. Wilz (eds.), *Organisationssoziologie. Formalität und Informalität in Organisationen*, Wiesbaden: Springer VS, 37–92.

Thorvaldsdottir, S., Patz, R., and Eckhard, S. (2021). International Bureaucracy and the United Nations System: Introduction, *International Review of Administrative Sciences 87* (4): 695–700. DOI: 10.1177/00208523211038730.

Trondal, J. (2011). Bureaucratic Structure and Administrative Behaviour: Lessons from International Bureaucracies, *West European Politics 34* (4): 795–818. DOI: 10.1080/01402382.2011.572392.

Urpelainen, J. (2012). Unilateral Influence on International Bureaucrats: An International Delegation Problem, *Journal of Conflict Resolution 56* (4): 704–735. DOI: 10.1177/0022002711431423.

Verhoest, K., Peters, B. G., Bouckaert, G., and Verschuere, B. (2004). The Study of Organisational Autonomy: A Conceptual Review, *Public Administration and Development 24* (2): 101–118.

Weinlich, S. (2014). *The UN Secretariat's Influence on the Evolution of Peacekeeping, Transformations of the State*, Basingstoke: Palgrave Macmillan.

Widerberg, O. and van Laerhoven, F. (2014). Measuring the Autonomous Influence of an International Bureaucracy: The Division for Sustainable Development, *International Environmental Agreements: Politics, Law and Economics* 14 (4): 303–327. DOI: 10.1007/s10784-014-9249-2.

Wilson, J. Q. (1989). *Bureaucracy: What Government Agencies Do and Why They Do It*, New York: Basic Books.

# 3

# The Evolution of International Environmental Bureaucracies

*How the Climate Secretariat Is Loosening Its Straitjacket*

THOMAS HICKMANN, OSCAR WIDERBERG, MARKUS LEDERER,
AND PHILIPP PATTBERG

## 3.1 Introduction

This chapter explores the growing influence of the Secretariat of the *United Nations Framework Convention on Climate Change* (UNFCCC) in contemporary global climate governance. Compared to other intergovernmental treaty secretariats, the political influence of the UNFCCC Secretariat has traditionally been considered rather limited. Most notably, Busch (2009: 245) characterized the role of the UNFCCC Secretariat as "Making a Living in a Straitjacket." However, we contend that the UNFCCC Secretariat has lately adopted a novel strategy to exert influence on the outcome of international climate negotiations and global climate policymaking by orchestrating the various climate initiatives undertaken by subnational and nonstate actors. Orchestration offers a "soft touch" approach and is an indirect mode of governance whereby a given agent interacts with intermediaries to influence a certain target group (Abbott and Snidal 2010; Abbott et al. 2015). Building upon this concept, we conceptualize the UNFCCC Secretariat as an international (environmental) bureaucracy that uses and works with cities as well as civil society groups, investors, and companies in order to aim at creating a momentum that nudges national governments to take more ambitious climate actions (Abbott and Bernstein 2015).

We perceive the UNFCCC Secretariat as an illustrative case for studying how international environmental bureaucracies can evolve from a rather low-key and servant-like secretariat to an actor in its own right – taking on the role of an orchestrator that seeks to shape policy outcomes through changing the behavior of others. Using orchestration as a conceptual lens, we identify new types of influence that were apparently not in the minds of those authors that studied the role and function of international bureaucracies as managers of global environmental problems about ten years ago (Biermann and Siebenhüner 2009). In particular, we recognize (i) awareness-raising, (ii) norm-building, and (iii) mobilization as forms of influence that the UNFCCC Secretariat exerts in global climate policymaking. This new way of how soft power is deployed underscores the increasingly proactive role of the UNFCCC Secretariat in the global response to climate change.

Empirically, this chapter zooms into three initiatives: the *Momentum for Change Initiative*, the *Lima–Paris Action Agenda (LPAA)*, and the *Non-state Actor Zone for Climate Action (NAZCA)*. These initiatives have been created to enhance the overall effectiveness of the global response to climate change and push the intergovernmental process forward by coordinating the myriad initiatives of subnational governments, nonprofit organizations, and business entities. This in turn has contributed to the shift away from a "regulatory" climate regime toward a "catalytic and facilitative" approach, in which subnational and nonstate actors play a much more prominent role (Hale 2016: 12). Thus, the main argument of our contribution is that orchestration entails indeed a specific form of influence, and although we cannot evaluate whether this will ultimately lead to more effective global climate policymaking, we show that the UNFCCC Secretariat is no longer a passive bystander but has adopted new roles and functions in the global endeavor to cope with climate change.

To advance our argument the chapter is structured as follows: Section 3.2 reviews the literature on influence exerted by international environmental bureaucracies. Then, we link this discussion to the concept of orchestration and sketch our methods of data collection. Section 3.3. provides a brief overview of the UNFCCC Secretariat and then focuses on the three initiatives in which the secretariat interacts with sub- and nonstate actors. Finally, Section 3.4 draws conclusions about the growing influence exerted by the UNFCCC Secretariat in global climate policymaking and points to some aspects that from our point of view merit attention in future research.

## 3.2 The Growing Influence of International (Environmental) Bureaucracies

Biermann and Siebenhüner (2009: 1) asked, "What is the role of international bureaucracies in world politics?" The two scholars argued that the literature in the fields of international relations management and legal studies underestimates the degree and variance of influence that these institutions have in global affairs. Influence is defined by Biermann et al. (2009: 40) who follow *Webster's Dictionary* as "the bringing about of an effect ... by a gradual process; controlling power quietly exerted." They deliberately do not speak of power as the connotation of coerciveness would be inherent, although the association to soft power (Nye 2004) is quite obvious. They further differentiate the observable effects that bureaucratic agents can bring about on the levels of *output*, *outcome*, and *impact*.

The following sections mainly focus on *outcome* since effects on the output level are easy to achieve but do not change much and measuring the (environmental) impact of administrative practices is already difficult within a domestic setting while hardly possible in global multilevel settings. Analyzing influence on the level of outcomes thus implies studying how the behavior of other actors has been targeted and has eventually changed, for example, in the sense of the targeted actor doing something different and becoming more compliant to an international rule-setting (Biermann et al. 2009: 43). In the analysis in this chapter, we will build on these conceptual ideas and analyze what kind of observable outcomes the UNFCCC Secretariat can achieve in terms of changing the behavior of actors softly and indirectly through orchestration techniques.

Any exercise in assessing the influence of intergovernmental public agencies, be it the bureaucracies of international organizations or small treaty secretariats, faces the well-established (neo)realist criticism that such effects are at best intervening factors and that the true power lies with nation-states and their respective central governments (Krasner 1986; Mearsheimer 1994). Hence, international bureaucracies might be able to facilitate or provide technical assistance and services to national governments but will in the end anticipate the preferences of the most powerful national governments (Drezner 2007). However, we would argue not only that the power of international organizations is gradually growing (Barnett and Finnemore 2004) but also that international secretariats have lately adopted more authoritative functions in global policymaking and gained increasing autonomy vis-à-vis their principals (Bauer and Ege 2016). International bureaucracies are capable of not just providing more and more output through setting up rules and procedures. They actually provide goods and services and influence other actors also on the outcome level. In fact, they do so independently from the broader development within the institutional structure they are part of and embedded in. We thus claim that international bureaucracies are distinct and partially influential actors that exercise important policymaking tasks (Eckhard and Ege 2016). While we see this as a broader phenomenon of global politics, it is particularly prevalent for international environmental bureaucracies (e.g., Hickmann and Elsässer 2020; Manulak 2017).

Studies in this field have advanced our knowledge on the role and function of international bureaucracies by looking at the specific mechanisms that bureaucracies have at their disposal to provide meaningful outcome and thus have achieved a certain level of influence (e.g., Jörgens et al. 2017). In line with these scholars, we investigate the new influence of international environmental bureaucracies and the attempt to use subnational and nonstate actors for achieving progress in the international environmental negotiations. The UNFCCC Secretariat does so through (i) awareness-raising, (ii) norm-building, and (iii) mobilization and we claim this can be best understood as elements of orchestration.

## 3.3 Taking Influence on Global Environmental Policymaking through Orchestration

Orchestration is a mode of governance that has gained increasing prominence in the disciplines of international relations and international law since it was popularized by Abbott and Snidal in 2009. These two scholars argued that a new regulatory structure started to emerge from the ashes of a failed "old governance system," in which subnational and nonstate actors take a more pronounced position by creating innovative transnational norms for regulating businesses (Abbott and Snidal 2009). Private and voluntary standards (Abbott 2012; Green 2013; Hickmann 2017b) are changing the global system of rules and norms away from traditional international governance through multilateral treaty-making under UN auspices toward a more heterogeneous, hybrid, and polycentric structure (Abbott, Green, and Keohane 2016; Biermann et al. 2009; Bulkeley et al. 2014; Hickmann 2016, 2017a; Jordan et al. 2015; Keohane and Victor 2011; Ostrom 2010). International organizations could use these new transnational institutions to "attain transnational regulatory goals that are not achievable through domestic or international Old Governance" (Abbott and Snidal 2009: 564).

Taking up this thread, Hale and Roger (2014: 60–61) defined orchestration as "a process whereby states or intergovernmental organizations initiate, guide, broaden, and strengthen transnational governance by non-state and/or sub-state actors." Hence, orchestration moves beyond the classical sender–receiver model of other governance approaches. It rather follows a so-called O–I–T model, in which an *O*rchestrator uses an *I*ntermediary to influence a certain *T*arget group. International organizations and their bureaucracies can in principle make use of various intermediaries, such as transnational networks, nongovernmental organizations, or public–private partnerships (Abbott et al. 2015: 6). Orchestrators have a wide range of techniques at their disposal to influence the intermediary, including direct assistance, endorsement, and coordination.

In theory, the orchestrator can thus choose to *manage* or *bypass* its targets that are in this study conceived of as nation-states. More precisely, orchestrators manage states when they enlist "intermediaries to shape state preferences, beliefs and behavior in ways that enhance state consent to and compliance with IGO [*international governmental organization*] goals policies and rules" (Abbott et al. 2015: 11). Orchestrators bypass nation-states when they approach and enlist intermediaries directly, to supply some kind of a common pool resource or public common. In the case of international organizations or bureaucracies as orchestrators, these can hence fulfill their purpose without needing "time-consuming, high-level political approval" (Abbott and Snidal 2009: 564).

Orchestration techniques employed by international organizations or bureaucracies as a mode of governance represent a shift in direction of authority, in particular if one adopts a traditional *principal–agent* perspective that centers around delegation of authority *from* a principal *to* an agent (Hawkins et al. 2006a). In the case of the UNFCCC Secretariat, nation-states can be considered the principals and the secretariat the agent. Any deviance by the agent from the mandate it has received from the principals constitutes *agency slack*, which generally means minimizing the effort by the agent to fulfill its primary mission (shirking) or taking actions that are contrary to the principals' desired policy direction (slippage) (Hawkins et al. 2006b: 8). Moreover, managing and bypassing nation-states would arguably be beyond what can be considered the *discretionary space* that an agent may be given within the mandate by the principals to accomplish certain tasks.

Whether and to what extent the UNFCCC Secretariat has been engaging in such shirking or slippage actions by orchestrating subnational and nonstate initiatives is an open question. In the past few years, the UNFCCC Secretariat has been described as a potential candidate for orchestrating climate governance at various stages of the policy cycle. Hale and Roger (2014: 80) argue that "it is possible to imagine the UNFCCC taking a more 'orchestrative' role than it does today." Yet they also acknowledge that "[w]hile it is unlikely to adopt and support, much less launch, particular transnational initiatives, it … could nonetheless be used as a forum for information-sharing, standard-setting, and accountability for transnational initiatives, and for focusing expectations on such practices" (Hale and Roger 2014: 80). A similar argument has been made by Chan et al. (2015: 470) who suggest that

[t]he UNFCCC secretariat on its own lacks the necessary resources, the mandate to ensure nonstate accountability, and the connections with nonstate actors to manage a comprehensive framework, hamstringing its operational effectiveness and experimental and catalytic abilities. At the same time, the secretariat has an important role to play. With universal membership, the UNFCCC provides the secretariat great legitimacy to convene and orchestrate nonstate initiatives in pursuit of public goals.

Thus, these scholars argue that the UNFCCC Secretariat could likely adopt an important role as an orchestrator, while it lacks the mandate, budget, and capacity. Nevertheless, even within these constraints, the suggestions by Hale and Roger as well as Chan et al. go far beyond previous understandings of the secretariat as "Making a living in a straitjacket" (Busch 2009: 245). While previous research has by and large discussed the potential of the UNFCCC Secretariat as an orchestrator, we move toward empirical assessment. In order to understand the mandate under which the secretariat operates, the following sections explore three illustrative examples of how the UNFCCC Secretariat has expanded its influence on the outcome level on global climate policymaking by using orchestration as a mode of governance, in particular after the failure at the 2009 climate summit in Copenhagen to reach a new international climate treaty.

The analysis relies on three sources of information: (i) an extensive desk study of existing scholarly work on the role and function of international bureaucracies and the UNFCCC Secretariat; (ii) a systematic content analysis of official documents, online material, and "grey" literature on the different initiatives in which the UNFCCC Secretariat interacts with sub- and nonstate actors; and (iii) a series of seventeen semistructured expert interviews with staff members of the UNFCCC Secretariat as well as representatives of different subnational bodies and nongovernmental organizations.

### 3.4 Studying the Influence of the UNFCCC Secretariat

The origins of the UNFCCC Secretariat date back to early 1991 when the then Secretary-General of the United Nations, Javier Pérez de Cuéllar, assigned a higher official in the *United Nations Conference on Trade and Development* with the task of building up a team of about a dozen people to support the intergovernmental negotiations that led to the adoption of the UNFCCC in 1992 (Yamin and Depledge 2004: 487). After a steady increase in tasks and personnel over the past two decades, the UNFCCC Secretariat now employs about 500 people (including both higher level employees and administrative posts) and possesses a yearly budget of approximately USD 90 million (UNFCCC 2017g).

Several scholars have addressed the role and functions of the UNFCCC Secretariat in global climate policymaking (Bauer, Busch, and Siebenhüner 2009; Busch 2009; Depledge 2005, 2007; Hickmann et al. 2021; Jörgens, Kolleck, and Saerbeck 2016; Michaelowa and Michaelowa 2016; Yamin and Depledge 2004). These scholars concede that the UNFCCC Secretariat maintains an important position with regard to the organization of the international climate negotiations and in supporting the various associated institutions and subbodies. However, most scholars have considered the broader political influence exerted by the UNFCCC Secretariat on other agents as rather low and have not claimed that a real influence on behavioral change is really discernable. Only the most recent accounts hold that the secretariat has developed an observable influence on global climate policymaking (Jörgens, Kolleck, and Saerbeck 2016; Michaelowa and Michaelowa 2016).

As previously indicated, Busch (2009: 251) most prominently claims in his case study that "[t]he climate secretariat is a 'technocratic bureaucracy' that has not had any autonomous political influence." He identifies the particular problem structure of the policy domain of climate change as a main reason for the limited leeway of the UNFCCC Secretariat and argues that the UNFCCC Secretariat has been put into a "straitjacket [which] reduces the potential for the climate secretariat to effectively exploit its key position and to have autonomous influence" (Busch 2009: 256). However, we put forward the argument that the UNFCCC Secretariat has lately been involved in a number of initiatives that seek to incorporate subnational and nonstate actors more directly and this potentially allows new forms of leverage.

In the following analysis, we first outline interactions between the secretariat and subnational and nonstate actors within the intergovernmental process. Then, we discuss three initiatives of the UNFCCC Secretariat's engagement with subnational and nonstate actors (i.e., the *Momentum for Change Initiative*, the *LPAA*, and *NAZCA*). We do not claim that the UNFCCC Secretariat has substantially altered the international landscape of climate politics and it can be questioned whether the described activities will have a discernable impact. Yet influence in the sense of changing the behavior of actors through orchestration is clearly visible.

## *The UNFCCC Secretariat and Subnational and Nonstate Actors*

The UNFCCC Secretariat has a long tradition in working together with nongovernmental organizations. Since the first *Conference of the Parties* (COP) to the UNFCCC, in 1995, the UNFCCC Secretariat has coordinated the participation of the constantly growing number of observer organizations in the international climate change conferences and the various accompanying events. Moreover, it has taken responsibility of the administration of side events conducted by all kinds of nongovernmental organizations. By this means, the UNFCCC Secretariat creates a forum for these actors and facilitates the informal exchange between different stakeholders that provide input to the intergovernmental negotiations and stimulate debates on a great variety of topics connected to the issue of climate change (Schroeder and Lovell 2012). However, these activities do not necessarily have a direct influence on national governments.

The COP17 held in Durban in 2011 and the *Ad Hoc Working Group on the Durban Platform for Enhanced Action* provided an opportunity for the UNFCCC Secretariat to interact with subnational and nonstate actors under an expanded mandate (UNFCCC 2011b). This subsidiary body of the UNFCCC was structured according to two different *workstreams*. Under the first workstream (WS1), nation-states agreed to negotiate a new legally binding agreement applicable to all parties to the UNFCCC, which led to the adoption of the *Paris Agreement* at COP21 in 2015 (UNFCCC 2015b). The second workstream (WS2) aimed to reduce the gap between the current efforts to reduce global greenhouse gas emissions and the goal of limiting global warming within the range of 1.5°C to 2°C. It established a framework for concrete short- to medium-term mitigation actions (up to 2020) to ensure the highest efforts by all nation-states as well as other relevant actors, including subnational governments, civil society groups, and private companies.

The UNFCCC Secretariat had two important tasks under WS2 relating to subnational and nonstate climate action. First, it organized a number of workshops and conducted so-called *Technical Expert Meetings* involving both public bodies and private/business actors "to share policies, practices and technologies and address the necessary finance, technology and capacity building, with a special focus on actions with high mitigation potential" (UNFCCC 2014c: 6). In this context, the secretariat was asked to synthesize the outcomes of the events into reports and summaries for policymakers (Hermwille et al. 2015: 15–16). Second, it was asked to compile information on action that could enhance the mitigation ambitions of governments, including many hybrid and private initiatives, into *Technical Papers*. These initiatives acknowledged the role the Secretariat could play in helping parties to support such "cooperative initiatives" (Widerberg and Pattberg 2015). Moreover, the secretariat launched a database to gather information on the various so-called *International Cooperative Initiatives* undertaken by national or subnational governments and all types of nongovernmental organizations (UNFCCC 2017f).

While the actions undertaken by the UNFCCC Secretariat under WS2 could largely be considered to fall within its mandate to facilitate the international negotiations, the remainder of this analysis focuses on initiatives of the secretariat to take a stronger impact on global climate policymaking by incorporating subnational and nonstate actors more directly into a policy dialogue. In these initiatives, sub- and nonstate entities are not merely observers of the international negotiations but have become actors that implement climate projects by themselves. According to a staff member of the UNFCCC Secretariat, the new strategy pursued by the executive secretary was to reach beyond the "usual conference hoppers."[1] In these initiatives, we recognize (i) awareness-raising, (ii) norm-building, and (iii) mobilization as new forms of influence exerted by the UNFCCC Secretariat that changed the behavior of these actors.

### *Awareness-Rising: The Momentum for Change Initiative*

An early initiative that was spearheaded by the UNFCCC Secretariat is the *Momentum for Change Initiative* (UNFCCC 2011a). It was officially presented to the public in 2011 to "get in a sense of optimism" into the negotiations and to "showcase climate solutions."[2] The initiative was not directly funded through the UNFCCC Secretariat's budget, as such activities would not have been

---

[1] Interview with Ian Ponce, Programme Officer with the UNFCCC Secretariat in the area of Strategy and Relationship Management, October 6, 2016, in Bonn, Germany.
[2] Interview with Luis Dávila, Programme Officer with the UNFCCC Secretariat in the Momentum for Change Initiative, October 6, 2016, in Bonn, Germany.

covered by its mandate. Instead, the team led by Christiana Figueres started to contact institutions like the *Bill and Melinda Gates Foundation*, the *Women in Sustainability, Environment and Renewable Energy Initiative*, the *World Economic Forum*, the *Rockefeller Foundation*, and the *Global e-Sustainability Initiative* to gather funds. In this way, national governments could not officially object to this outreach campaign and in the end even welcomed the process, a fact that surprised some of those who were involved in the project from the beginning.

The proclaimed goal of this initiative is "to shine a light on the enormous groundswell of activities underway across the globe that are moving the world towards a highly resilient, low-carbon future" (UNFCCC 2017d). To reach this goal, the initiative recognizes a number of so-called *Lighthouse Activities*, which are described as innovative and transformative solutions of civil society organizations and business associations or firms addressing both climate-related aspects and wider economic, social, and environmental challenges in a given geographical area. According to the initiative's webpage, these particular activities are practical, scalable, and replicable examples of what societal actors are doing to cope with the problem of climate change.

Since 2012, the initiative confers the *Momentum for Change Awards* to particularly successful climate change mitigation or adaptation projects conducted by nonstate actors from around the world. The initiative has four different focus areas: (i) Urban Poor: recognizing climate actions that improve the lives of impoverished people in urban areas, supported by the *Bill and Melinda Gates Foundation*; (ii) Women for Results: recognizing critical leadership and participation of women, implemented together with the *Women in Sustainability, Environment and Renewable Energy Initiative*; (iii) Financing for Climate Friendly Investment: recognizing successful and innovative climate-smart activities, in cooperation with the *World Economic Forum*; and (iv) ICT Solutions: recognizing climate-relevant projects in the field of information and communication technology, carried out with the *Global e-Sustainability Initiative*.

In the past few years, the UNFCCC Secretariat has put considerable efforts into the development of this initiative and established numerous partnerships with the private sector to engage in mutually beneficial collaborative interactions in order to raise public awareness on climate actions taking place on the ground (e.g., UNFCCC 2012, 2014a, 2015a, 2017e). In late 2016, four staff members were working on this initiative.[3] Among insiders, it has also been described as the "pet initiative" of Christiana Figueres, and when asked how the project evolved,

---

[3] Interview with Luis Dávila, Programme Officer with the UNFCCC Secretariat in the Momentum for Change Initiative, October 6, 2016, in Bonn, Germany.

one of the responsible officials simply answered "only the sky is the limit."[4] This underscores that the UNFCCC Secretariat has acquired a new form of influence in global climate policymaking by using nonstate actors to raise awareness on the issue of climate change and push national governments to take a more ambitious stance on climate change.

### *Norm-Building: The LPAA*

In the run-up to Paris, the LPAA was launched in December 2014, during COP20 in Lima. Its primary goal was to boost the positive momentum created by various conferences organized by the United Nations Secretary-General's Office throughout 2014 that targeted sub- and nonstate actors. The LPAA was jointly launched by the Peruvian and French COP presidencies, the Executive Office of the United Nations Secretary-General, and the UNFCCC Secretariat (United Nations 2015). The common intention of these four actors was to accelerate the growing engagement of all parts of society in climate action and to build concrete, ambitious, and lasting initiatives that will help reduce global greenhouse gas emissions and promote measures to better adapt to the various adverse effects associated with the problem of climate change (Widerberg 2017).

While the UNFCCC Secretariat had only a relatively small part in the launch and the run-up to the initiative, it adopted a substantial role throughout 2015. Prior to COP21, for instance, it published a policy paper that called for further evolution of the initiative together with the Peruvian and French governments as well as the Executive Office of the United Nations Secretary-General (UNFCCC 2017b). Moreover, the secretariat supervised the initiative and occupied two seats in its steering committee that is responsible for the initiative's strategic development and implementation. It did not, however, go as far as some of the other partners in the LPAA that provided temporary administrative bodies and acted as conveners for new initiatives to be launched in Paris (Widerberg 2017).

Yet the LPAA allowed the UNFCCC Secretariat to explore new territory and acquire new forms of influence in global climate policymaking by involving nation-states, cities, regions, and other subnational entities, international organizations, civil society, indigenous peoples, women, youth, academic institutions, and companies and investors to build a norm that a new climate treaty should be adopted in Paris.[5] The LPAA was designed to catalyze climate action in the short

---

[4] Interview with a former staff member of the UNFCCC Secretariat who wished to remain anonymous, October 7, 2016, in Bonn, Germany.
[5] Interview with Ian Ponce, Programme Officer with the UNFCCC Secretariat in the area of Strategy and Relationship Management, October 6, 2016, in Bonn, Germany.

term, especially by building momentum toward the end of 2015 and support the negotiation of a new agreement, as well as in the long term, before and after 2020 when the *Paris Agreement* took effect.

## *Mobilization: NAZCA*

The third initiative concerns the secretariat's engagement in the launch and maintenance of NAZCA. In 2014, the UNFCCC Secretariat supported the Peruvian government in the launch of NAZCA, which is an online platform to coordinate the various climate-related activities of nonstate actors and to register their individual commitments (Chan et al. 2015: 468). The aim of this initiative is to improve the visibility of climate actions by subnational and nonstate actors (UNFCCC 2017a). In particular, NAZCA should demonstrate how the momentum for climate action is rising and showcase the "extraordinary range of game-changing actions being undertaken by thousands of cities, investors and corporations" (UNFCCC 2014b).

The "theory of change" is that national governments would be more inclined to reach an ambitious agreement in the Paris meeting if they knew that their constituencies also favored strong climate action (Widerberg 2017). Jacobs (2016: 322), for instance, argues that "[b]y orchestrating the narratives of science and economics to demand strong climate action, and organising the business community, NGOs and many others in support of a strong agreement, it was civil society that pressured governments into the positions that made the final negotiations possible." NAZCA draws on data from established and credible sources with a strong record of reporting and tracking progress, such as the *Carbon Disclosure Project* and the *carbonn* [sic] *Climate Registry* (Widerberg and Stripple 2016). In 2017, the platform comprised 12,549 total commitments, out of which 2,508 have been announced by cities, 209 by regions, 2,138 by companies, 479 by investors, and 238 by civil society organizations (UNFCCC 2017c).

In addition to running the platform, the UNFCCC Secretariat regularly carries out consultations with stakeholders on potential improvements. This also indicates that the UNFCCC Secretariat has recently expanded its role and attained a new form of influence by actively working together with actors other than national governments in the pursuit of the general aim of mobilization of global mitigation and adaptation actions. In this context, a staff member of the UNFCCC Secretariat noticed that NAZCA also contributed to the formal inclusion of sub- and nonstate actors into the Paris Agreement "shining a light on the numerous existing successful climate actions."[6]

---

[6] Interview with Ian Ponce, Programme Officer with the UNFCCC Secretariat in the area of Strategy and Relationship Management, October 6, 2016, in Bonn, Germany.

## 3.5 Conclusions

This chapter explored the growing influence of the UNFCCC Secretariat in contemporary global climate governance. Based on the previous analysis, we put forward the argument that the secretariat has lately adopted a novel strategy to exert influence on the outcome level of climate policymaking by orchestrating the various climate initiatives undertaken by subnational and nonstate actors.

We particularly recognize (i) awareness-raising, (ii) norm-building, and (iii) mobilization as forms of influence that the UNFCCC Secretariat exerts in the global response to climate change. In the Momentum for Change initiative, the secretariat has used nonstate actors to raise awareness on the issue of climate change and push national governments to take a more ambitious stance on climate change. In the LPAA, the secretariat acquired new forms of influence by involving the parties as well as all sorts of subnational and nonstate actors to build a norm that a new climate treaty should be adopted in Paris. Finally, the secretariat put considerable efforts into the launch and maintenance of the NAZCA platform to accelerate and mobilize the global mitigation and adaptation ambition. These findings suggest that the UNFCCC Secretariat has found new ways to exert influence on the intergovernmental process by interacting with sub- and nonstate actors with the overall aim of inducing national governments to adopt more progressive climate targets.

In addition, the UNFCCC Secretariat used the different initiatives for a new communication strategy reaching out to the media and certain celebrities. This is in line with what Jörgens et al. (2017) recently termed an "attention-seeking bureaucracy." In other words, the UNFCCC Secretariat essentially operated according to the principle *Do Good and Make It Known*. Policywise, the overall objective of these initiatives is to reinvigorate the global endeavor to address climate change by emphasizing pioneering climate initiatives of cities and their networks, civil society groups, nonprofit entities, and private companies as well as business associations. In this way, momentum shall be built up for an increased level of ambition to address climate change. The analysis hence suggests that the UNFCCC Secretariat can no longer be adequately described as a purely technocratic international environmental bureaucracy (Hickmann and Elsässer 2020; Hickmann et al. 2021). Instead, in this chapter we put forward the argument that the secretariat influences not only the output but also the outcome level in the field of global climate politics.

The UNFCCC Secretariat took a certain window of opportunity and involved subnational and nonstate actors as a novel strategy, influencing them to raise the global level of ambition to address climate change. Through its outreach strategy and policy dialogue with actors other than national governments, the secretariat provided impetus for a variety of climate-related projects in all parts of the world carried out by subnational governments, nonprofit entities, and private businesses.

These findings suggest that the UNFCCC Secretariat could loosen its straitjacket and in recent years considerably expand its political influence in the global response to climate change.

## References

Abbott, K. W. (2012). The Transnational Regime Complex for Climate Change, *Environment and Planning C: Government and Policy* 30 (4): 571–590. DOI: 10.1068/C11127.

Abbott, K. W. and Bernstein, S. (2015). The High-Level Political Forum on Sustainable Development: Orchestration by Default and Design, *Global Policy* 6 (3): 222–233.

Abbott, K. W. and Snidal, D. (2009). Strengthening International Regulation through Transnational New Governance: Overcoming the Orchestration Deficit, *Vanderbilt Journal of Transnational Law* 42 (2): 501–578.

Abbott, K. W. and Snidal, D. (2010). International Regulation without International Government: Improving IO Performance through Orchestration. *The Review of International Organizations* 5 (3): 315–344.

Abbott, K. W., Green, J. F., and Keohane, R. O. (2016). Organizational Ecology and Institutional Change in Global Governance, *International Organization* 70 (2): 247–277.

Abbott, K. W., Genschel, P., Snidal, D., and Zangl, B. (eds.) (2015). *International Organizations as Orchestrators*, Cambridge: Cambridge University Press.

Barnett, M. and Finnemore, M. (2004). *Rules for the World: International Organizations in Global Politics*, Ithaca, NY: Cornell University Press.

Bauer, M. W. and Ege, J. (2016). Bureaucratic Autonomy of International Organizations' Secretariats, *Journal of European Public Policy* 23 (7): 1019–1037. DOI: 10.1080/13501763.2016.1162833.

Bauer, S., Busch, P.-O., and Siebenhüner, B. (2009). Treaty Secretariats in Global Environmental Governance. In F. Biermann, B. Siebenhüner, and A. Schreyögg (eds.), *International Organizations in Global Environmental Governance*, London: Routledge, 174–192.

Biermann, F. and Siebenhüner, B. (2009). The Role and Relevance of International Bureaucracies: Setting the Stage. In F. Biermann and B. Siebenhüner (eds.), *Managers of Global Change: The Influence of International Environmental Bureaucracies*, Cambridge, MA: MIT Press, 1–14.

Biermann, F., Pattberg, P., van Asselt, H., and Zelli, F. (2009). The Fragmentation of Global Governance Architectures: A Framework for Analysis, *Global Environmental Politics* 9 (4): 14–40. http://muse.jhu.edu/content/z3950/journals/global_environmental_politics/v009/9.4.biermann.html

Biermann, F., Siebenhüner, B., Bauer, S., et al. (2009). Studying the Influence of International Bureaucracies: A Conceptual Framework. In F. Biermann and B. Siebenhüner (eds.), *Managers of Global Change: The Influence of International Environmental Bureaucracies*, Cambridge, MA: MIT Press, 37–74.

Bulkeley, H., Andonova, L., Betsill, M., et al. (2014). *Transnational Climate Change Governance*, New York: Cambridge University Press.

Busch, P.-O. (2009). The Climate Secretariat: Making a Living in a Straitjacket. In F. Biermann and B. Siebenhüner (eds.), *Managers of Global Change: The Influence of International Environmental Bureaucracies*, Cambridge, MA: MIT Press, 245–264.

Chan, S., van Asselt, H, Hale, T., et al. (2015). Reinvigorating International Climate Policy: A Comprehensive Framework for Effective Nonstate Action, *Global Policy* 6 (4): 466–473.

Depledge, J. (2005). *The Organization of International Negotiations: Constructing the Climate Change Regime*, London: Earthscan.

Depledge, J. (2007). A Special Relationship: Chairpersons and the Secretariat in the Climate Change Negotiations, *Global Environmental Politics* 7 (1): 45–68. DOI: 10.1162/glep.2007.7.1.45.

Drezner, D. W. (2007). *All Politics Is Global: Explaining International Regulatory Regimes*, Princeton: Princeton University Press.

Eckhard, S. and Ege, J. (2016). International Bureaucracies and Their Influence on Policy-Making: A Review of Empirical Evidence, *Journal of European Public Policy* 23 (7): 960–978.

Green, J. F. (2013). Order Out of Chaos: Public and Private Rules for Managing Carbon, *Global Environmental Politics* 13 (2): 1–25. DOI: 10.1162/Glep_a_00164.

Hale, T. (2016). "All Hands on Deck": The Paris Agreement and Nonstate Climate Action, *Global Environmental Politics* 16 (3): 12–22.

Hale, T and Roger, C. (2014). Orchestration and Transnational Climate Governance, *Review of International Organizations* 9 (1): 59–82.

Hawkins, D. G., Lake, D. A., Nielson, D. L., and Tierney, M. J. (eds.) (2006a). *Delegation and Agency in International Organizations*, Cambridge: Cambridge University Press.

Hawkins, D. G., Lake, D. A., Nielson, D. L., and Tierney, M. J. (2006b). Delegation under Anarchy: States, International Organizations, and Principal-Agent Theory. In D. G. Hawkins, D. A. Lake, D. L. Nielson, and M. J. Tierney (eds.), *Delegation and Agency in International Organizations*, Cambridge: Cambridge University Press, 3–38.

Hermwille, L., Obergassel, W., Ott, H. E., and Beuermann, C. (2015). UNFCCC before and after Paris: What's Necessary for an Effective Climate Regime? *Climate Policy* 17 (2): 150–170.

Hickmann, T. (2016). *Rethinking Authority in Global Climate Governance: How Transnational Climate Initiatives Relate to the International Climate Regime*, London: Routledge.

Hickmann, T. (2017a). The Reconfiguration of Authority in Global Climate Governance, *International Studies Review* 19 (3): 430–451.

Hickmann, T. (2017b). Voluntary Global Business Initiatives and the International Climate Negotiations: A Case Study of the Greenhouse Gas Protocol, *Journal of Cleaner Production* 169: 94–104.

Hickmann, T. and Elsässer, J. (2020). New Alliances in Global Environmental Governance: How Intergovernmental Treaty Secretariats Interact with Non-state Actors to Address Transboundary Environmental Problems, *International Environmental Agreements: Politics, Law and Economics* 20 (3): 459–481.

Hickmann, T., Widerberg, O., Lederer, M., and Pattberg, P. (2021). The United Nations Framework Convention on Climate Change Secretariat as an Orchestrator in Global Climate Policymaking, *International Review of Administrative Sciences* 87 (1): 21–38.

Jacobs, M. (2016). High Pressure for Low Emissions: How Civil Society Created the Paris Climate Agreement, *Juncture* 22 (4): 314–323.

Jordan, A. J., Huitema, D., Hildén, M., et al. (2015). Emergence of Polycentric Climate Governance and Its Future Prospects, *Nature Climate Change* 5 (11): 977–982.

Jörgens, H., Kolleck, N., and Saerbeck, B. (2016). Exploring the Hidden Influence of International Treaty Secretariats: Using Social Network Analysis to Analyse the Twitter Debate on the "Lima Work Programme on Gender," *Journal of European Public Policy* 23 (7): 979–998.

Jörgens, H., Kolleck, N., Saerbeck, B., and Well, M. (2017). Orchestrating (Bio-)Diversity: The Secretariat of the Convention of Biological Diversity as an Attention-Seeking Bureaucracy. In M. W. Bauer, C. Knill, and S. Eckhard (eds.), *International Bureaucracy: Challenges and Lessons for Public Administration Research*, Basingstoke: Palgrave Macmillan, 73–95.

Keohane, R. O. and Victor, D. G. (2011). The Regime Complex for Climate Change, *Perspectives on Politics* 9 (1): 7–23.

Krasner, S. D. (1986). Structural Causes and Regime Consequences: Regimes as Intervening Variables. In S. D. Krasner (ed.), *International Regimes*, Ithaca, NY: Cornell University Press, 1–21.

Manulak, M. W. (2017). Leading by Design: Informal Influence and International Secretariats, *The Review of International Organizations* 12 (4): 497–522. DOI: 10.1007/s11558-016-9245-0.

Mearsheimer, J. J. (1994). The False Promise of International Institutions, *International Security* 19: 5–49.

Michaelowa, A. and Michaelowa, K. (2016). The Growing Influence of the UNFCCC Secretariat on the Clean Development Mechanism, *International Environmental Agreements: Politics, Law and Economics* (online first): 1–23.

Nye, J. S. (2004). *Soft Power: The Means to Success in World Politics*, Cambridge: Perseus Book Group.

Ostrom, E. (2010). Polycentric Systems for Coping with Collective Action and Global Environmental Change, *Global Environmental Change* 20 (4): 550–557. DOI: 10.1016/j.gloenvcha.2010.07.004.

Schroeder, H. and Lovell, H. (2012). The Role of Non-nation-state Actors and Side Events in the International Climate Negotiations, *Climate Policy* 12 (1): 23–37. DOI: 10.1080/14693062.2011.579328.

UNFCCC (2011a). Momentum for Change: Launch Report. https://unfccc.int/files/secretariat/momentum_for_change/application/pdf/mfc_launch_report1.pdf

UNFCCC (2011b). Report of the Conference of the Parties on Its Seventeenth Session, held in Durban from November 28 to December 11, 2011. Addendum: Part Two: Action Taken by the Conference of the Parties at Its Seventeenth Session, FCCC/CP/2011/9/Add.2, Bonn: UNFCCC Secretariat.

UNFCCC (2012). Momentum for Change in 2012: Momentum for Change – Change for Good. https://unfccc.int/files/secretariat/momentum_for_change/application/pdf/mfc_report.pdf

UNFCCC (2014a). Momentum for Change: Annual Report 2013. https://unfccc.int/files/secretariat/momentum_for_change/application/pdf/unfccc_mfc_annual_report_2013.pdf

UNFCCC (2014b). New Portal Highlights City and Private Sector Climate Action, Press Release of the UNFCCC Secretariat, December 11, 2014. https://unfccc.int/news/new-portal-highlights-city-and-private-sector-climate-action

UNFCCC (2014c). Report of the Ad Hoc Working Group on the Durban Platform for Enhanced Action on the Third Part of Its Second Session, held in Warsaw from November 12 to 23, 2013, FCCC/ADP/2013/3, Bonn: UNFCCC Secretariat.

UNFCCC (2015a). Momentum for Change: 2015 Lighthouse Activities. https://unfccc.int/news/2015-momentum-for-change-lighthouse-activities

UNFCCC (2015b). Report of the Conference of the Parties on Its Twenty-First Session, held in Paris from November 30 to December 13, 2015. Addendum: Part Two: Action Taken by the Conference of the Parties at Its Twenty-First Session, Bonn: UNFCCC Secretariat.

UNFCCC (2017a). About NAZCA. http://climateaction.unfccc.int/about

UNFCCC (2017b). About the Lima-Paris Action Agenda. http://newsroom.unfccc.int/lpaa/about/

UNFCCC (2017c). The Lima-Paris Action Agenda: Cooperative Initiatives. http://climateaction.unfccc.int/

UNFCCC (2017d). Momentum for Change. http://unfccc.int/secretariat/momentum_for_change/items/6214.php

UNFCCC (2017e). Momentum for Change: Lighthouse Activities. http://momentum.unfccc.int/

UNFCCC (2017f). Portal on Cooperative Initiatives. http://unfccc.int/focus/mitigation/items/7785.php#feedback

UNFCCC (2017g). The Secretariat: Who We Are. http://unfccc.int/secretariat/items/1629.php

United Nations (2015). Lima to Paris Action Agenda Driving Climate Action Forward. www.un.org/climatechange/blog/2015/06/lima-paris-action-agenda-driving-climate-action-forward/

Widerberg, O. (2017). The "Black Box" Problem of Orchestration: How to Evaluate the Performance of the Lima-Paris Action Agenda, *Environmental Politics* 26 (4): 715–737.

Widerberg, O. and Pattberg, P. (2015). International Cooperative Initiatives in Global Climate Governance: Raising the Ambition Level or Delegitimizing the UNFCCC?, *Global Policy* 6 (1): 45–56.

Widerberg, O. and Stripple, J. (2016). The Expanding Field of Cooperative Initiatives for Decarbonization: A Review of Five Databases, *Wiley Interdisciplinary Reviews: Climate Change* 7 (4): 486–500.

Yamin, F. and Depledge, J. (2004). *The International Climate Change Regime: A Guide to Rules, Institutions and Procedures*, Cambridge: Cambridge University Press.

# 4

# Environmental Treaty Secretariats as Attention-Seeking Bureaucracies

*The Climate and Biodiversity Secretariats' Role in International Public Policymaking*

MAREIKE WELL, HELGE JÖRGENS, BARBARA SAERBECK, AND NINA KOLLECK

## 4.1 Introduction

There is little doubt that international bureaucracies can be influential actors in world politics, as this volume emphasizes. The principal question asked by scholars of international public administration is "under which conditions and to what extent international [bureaucratic] influence emerges autonomously from political superiors" (Bauer and Ege 2016: 1021) and what the causal mechanisms are through which this influence occurs. In this chapter we argue that international bureaucracies turn into influential actors at the international level not by covertly attempting to influence international processes but by actively seeking the attention of states, which we illustrate with two case studies that zoom in on international climate and biodiversity politics. We start from a perspective of bureaucracies as institutions that have "a raison d'être and organizational and normative principles of its own" (Olsen 2006: 3) and are an essential element of a political system's decision-making capacity. This contrasts with a different perspective that regards bureaucracies primarily "as a rational tool for executing the commands of elected leaders" (Olsen 2006: 3). From this approach, autonomous bureaucratic influence occurs when bureaucrats hold policy-related preferences that deviate from those of their principals and exploit information asymmetries to shape political programs in accordance with their preferences (McCubbins, Noll, and Weingast 1987: 247). Scholars have focused primarily on the conditions under which unintended agency slack occurs and on the design of incentive structures to effectively control it (Hawkins et al. 2006). We suggest complementing the principal–agent perspective, which conceives of bureaucracies primarily as attention-avoiding organizations, with a public policy perspective that emphasizes the attention-seeking character of those bureaucracies, especially when involved predominantly in the formulation rather than the implementation of public policies. We build on a research tradition that is mainly concerned with policy outputs and bureaucracy's autonomous contribution to the problem-solving capacity of the political system as a whole, based on bureaucratic authority (Barnett and Finnemore 2004; Busch and Liese 2017). In this view, a certain degree of autonomy from governments and parliaments is seen as desirable and as a necessary precondition for bureaucracies to be able to "speak truth to power" and to fulfill their function as an independent political institution (Olsen 2006: 3). The bureaucratic authority of attention-seeking bureaucracies emphasizes an entrepreneurial stance and is not primarily delegated from their principals (Green 2014: 33; Well et al. 2020: 108).

Against this backdrop, we argue that international bureaucracies actively step into the limelight in order to feed their expert knowledge and policy preferences into the policymaking process of states. Our main argument is that international organizations and multilateral negotiations are limited not by a lack of information but by the capacity of negotiators to process and prioritize the enormous amount of information available. Thus, to influence international multilateral negotiation outcomes, bureaucracies need to attract the attention of state negotiators instead of withholding information from them. In order to illustrate and explore this attention-seeking character of public administrations, we focus on international treaty secretariats as a specific type of bureaucracy that is almost exclusively involved in the early stages of the policy process. Hence, we aim to identify the strategies international treaty secretariats as attention-seeking bureaucracies employ in the early stages of the policy cycle. We describe two potential pathways through which international treaty secretariats may attract the attention of the state parties to multilateral negotiations: (i) They can directly seek the attention of negotiators through close cooperation with, for example, the chairs or presidency of multilateral conferences and (ii) they can facilitate exchange and build up support for their problem definitions and policy recommendations outside of the official negotiation arenas.

The heuristic framework presented here not only is relevant for international bureaucracies but builds on recent research on the autonomy and influence of regulatory agencies in US policymaking (Carpenter 2001; Workman 2015). What this latter research and our approach have in common is a focus on the role of public administrations during the early stages of the policy process, particularly in processes of problem definition, agenda-setting, and policy formulation. With few or no implementation tasks, international treaty secretariats constitute ideal empirical cases for analyzing the mechanisms through which public administrations can have a (partially) autonomous impact on the definition of problems and the design of political programs. Our findings, therefore, will contribute to a recent body of literature studying the role of national as well as international public administrations as agenda-setters, policy entrepreneurs, or policy brokers at the interface of public policy analysis and public administration (e.g., Abbott et al. 2015; Jinnah 2014;

see also Chapters 3, 5, 8, and 9). In order to put our heuristic framework to an empirical test, we conducted two explorative case studies, in which we analyzed the attention-seeking behavior of the United Nations Framework Convention on Climate Change (UNFCCC) Secretariat and of the Convention on Biological Diversity (CBD) Secretariat. The case studies illuminate attention-seeking strategies of these secretariats during and between multilateral negotiations leading to the Paris Agreement on climate change in 2015. The next sections outline our heuristic framework, which is followed by an analysis of interaction strategies of the UNFCCC and CBD Secretariats with the parties and nonparty stakeholders of the respective conventions, using the heuristic framework. The approach is based on qualitative content analysis of interviews conducted with members of the secretariats and party representatives of the conventions and of documents that give insight into the interaction strategies of the secretariats, such as treaty texts, decisions, and reports. Apart from validating our heuristic framework, the findings of our case studies are relevant for the literature on influence and legitimacy in global governance as well as for current climate and biodiversity governance.

## 4.2 Heuristic Framework: International Secretariats as Attention-Seeking Bureaucracies

From its beginnings, public administration research has been concerned with the political control of bureaucracy and the degree to which bureaucracies can exert autonomous influence on politics and policies (Weber 2018). Normatively, this part of the public administration literature has debated "the appropriate range of discretion for bureaucrats in a democratic polity" (Frederickson et al. 2018: 12). Analytically, it has focused on whether and to what extent bureaucracies exert an autonomous influence on the formulation and the implementation of public policies. Contrasting with Wilson's (1887) normative postulate of a politics-administration dichotomy, which implies a strict separation of politics and bureaucracy, empirical analyses have shown that "political control over bureaucracy" and "bureaucratic control over policy" are just two sides of the same coin (Frederickson et al. 2018: 18–19). Alford et al. (2017: 752) therefore refer to the blurred line between the political and administrative realms as a "purple zone representing where the 'red' of political activity overlaps with the 'blue' of administration."

In the past two decades more and more scholars have started to treat international bureaucracies as autonomous and consequential actors and begun to empirically study their role in processes of international public policymaking (Biermann and Siebenhüner 2009; Hawkins et al. 2006; Reinalda and Verbeek 1998). So far, the most influential theoretical approaches for studying the (partially) autonomous role and influence of international bureaucracies are based on principal–agent models. Scholars adopting a principal–agent perspective argue that (international) bureaucracies hold preferences that deviate from those of their principals (i.e., states), thereby creating problems of oversight and control. Based on a distinction between "collective" and "multiple" principals (Nielson and Tierney 2003: 247), they outline different potential mechanisms through which "agency slack" may occur in international organizations or multilateral treaty systems. According to this view, bureaucracies become actors in their own right, operating "behind the scenes" without openly articulating their preferences and policy positions (Arrow 1985; Hawkins et al. 2006; Mathiason 2007). By withholding policy-relevant information from decision-makers, they may create or reinforce information asymmetries that in turn are the basis for their autonomous influence. They may also exploit constellations characterized by multiple principals by strategically aligning with selected states whose policy preferences are similar to those of the secretariat (Dijkstra 2017).

This chapter builds on these approaches by stressing the importance both of the possession of policy-relevant information and of strategies of alliance-building as the principal sources of autonomy and influence of international bureaucracies. However, our argument differs from these approaches in the way we conceptualize the exchange of policy-relevant information between international bureaucracies and negotiating parties. Our main argument is that international organizations and multilateral negotiations are not limited by a lack of information but by the capacity of negotiators to process and prioritize the enormous amount of information available. Thus, in order to influence negotiation outcomes, international secretariats need to attract the attention of state negotiators instead of withholding information from them. Unless they actively feed their policy-relevant information, problem definitions, and policy preferences into the multilateral negotiations, information provided by other, competing, organizations will prevail.

Consequently, the possession of policy- or process-relevant expert knowledge alone does not turn international bureaucracies into influential actors at the international level. There are two main reasons for this. First, in multilateral negotiations, the alleged informational advantage of treaty secretariats vis-à-vis the representatives of member states is often much smaller than what principal–agent models hold. National delegations typically consist of experienced negotiators with extensive substantial and procedural knowledge in the issue area under negotiation. They are part of a domestic ministerial bureaucracy that might be complemented with expert consultants, which gives them the same advantages of issue-specific expertise, procedural knowledge, and permanence that principal–agent theories see as the main advantage of bureaucratic agents (Barnett and Finnemore 2004; Biermann and Siebenhüner 2009). Thus, in multilateral treaty negotiations the principals of international bureaucracies are mostly themselves national bureaucrats rather than elected politicians, since the latter typically join multilateral conferences only at the final stage of negotiations (the "ministerial segments") (Depledge 2005: 194). There may even be tough competition between international and national bureaucrats when it comes to defining processes and policies. Who "wins" such a race for defining key policy and procedural choices may depend more on the individual capacities (such as staff time) national and international bureaucrats can invest into a given subject matter rather than on the availability of information. The dependence of national bureaucrats on the expert knowledge provided by secretariats is therefore limited and varies according to context (e.g., on the salience of the topic in national bureaucracy, which again determines how much staff time is allocated to a given matter). While information asymmetries may play an important role in on-the-ground operations of large international financial organizations like the International Monetary Fund (Cox and Jacobson 1973), they are less relevant for treaty secretariats with relatively small staff and few implementation tasks (Biermann and Siebenhüner 2009). Second, the early stages of the policymaking process – problem definition, agenda-setting, and policy formulation – are generally characterized by an excess rather than a lack of policy-relevant information, including diverging definitions of the underlying problem and competing proposals for feasible solutions (Workman 2015). Thus, even where information asymmetries between treaty secretariats and national delegation exist, they normally do not imply that negotiators feel dependent on the policy-relevant information held by secretariats and that they will actively seek this information. We therefore expect negotiators, especially those with strong domestic environmental bureaucracies, to recur to secretariat information, particularly in those cases where the secretariats build close relations to national delegations and actively promote this information. What counts is not only the quality of the information international treaty secretariats hold but the extent to which they manage to bring that information to the attention of the parties to multilateral negotiations.

We thus argue that, in order to become influential, international bureaucracies need to not only possess policy-relevant expert knowledge but also exploit the complex structures and actor constellations of international organizations or multilateral treaty systems in ways to make negotiators take notice and adopt some of the bureaucracy's policy positions (Jinnah 2014; see also Chapter 9). In other words, in order to influence the outcomes of multilateral negotiations, international secretariats need to actively and strategically seek to draw the attention of the negotiating parties to the problem definitions and policy prescriptions provided by the secretariat. Workman (2015: 3) developed this argument for the domestic policymaking process: "If the supply of information yields bureaucratic influence, then bureaucracies must be willing to be attention-seeking and attention-attracting organizations, rather than the backroom dealers of subsystem lore." In this chapter, we contend that this argument also holds for the international policy process.

We argue that international treaty secretariats may best be conceived of as attention-seeking bureaucracies. We develop a heuristic framework that includes two paths by which international secretariats may try to draw the attention of negotiating parties to their own problem definitions and policy recommendations: (i) They may try to supply policy-relevant information directly and from the inside by cooperating closely with a convention's chairpersons,[1] with its presidency, or with individual groups of countries, trying to use these as multipliers and (ii) they may attempt to build support for their preferred policy outputs by engaging with and communicatively connecting actors within the broader transnational policy network that surrounds multilateral negotiations in order to exert pressure on negotiators from the outside. In both cases, international treaty secretariats act as attention-seeking policy advocates rather than "undercover agents" who try to operate out of the negotiators' sight. The two strategies are not mutually exclusive and can be employed in combination. International treaty secretariats' attempts to influence international policy outputs may be motivated either by self-interest (Niskanen 2017) or by professional ethic reflecting what Barnett and Finnemore (2004: 72) describe as "conscientious experts trying to do their job." Whereas bureaucratic self-interest is usually linked to the survival of international bureaucracies and to the expansion of their mandates as well as their human and material resources, research on international environmental secretariats has shown that international bureaucrats often draw their motivation from deeply held policy beliefs combined with a professional dedication to the overall goals and objectives of their organization or treaty system (Bauer 2006; Depledge 2005: 65). Any combination of bureaucratic self-interest and professional ethic is also possible, for example, when the expansion of mandates is rooted in a treaty secretariat's holistic vision of a global policy problem and its potential solutions (Well et al. 2020).

## *Treaty Secretariats as Attention-Seeking Bureaucracies*

International secretariats are created to support governments in subsequent rounds of issue-specific negotiations within multilateral treaty regimes, which are mainly concerned with the adoption of new treaty provisions and the revision and refinement of existing ones (Gehring 2012: 51). In these treaty systems, responsibilities for implementation remain mostly at the national level. Thus, if international treaty secretariats wield autonomous influence, we can reasonably expect this influence

---

[1] These can be negotiations within the Conference of the Parties or the subsidiary bodies of the relevant conventions.

to occur primarily at the stages of problem definition, agenda-setting, and policy formulation. At these stages of the policy process, information asymmetries arguably play a secondary role. The limiting factor is not scarcity of policy-related knowledge but rather the limited capacity of decision-makers to pay attention to the abundance of problem- and policy-relevant information. As Workman (2015: 59) argues in his study on bureaucratic influence in US policymaking, "Information not provided by one entity will assuredly be supplied by another as organized interests, federal bureaucracies, and policy makers engage in the struggle to define the contours of debate." Instead, bureaucracies compete with other organizations in the provision of policy-relevant information to elected officials.

This constellation – multiple providers of policy-relevant information and a strictly limited capacity for attention on the side of decision-makers – is even more pronounced in multilateral treaty systems. Here, treaty secretariats compete with a multitude of domestic bureaucracies with strongly varying interests and preferences, other international organizations, scientific or nongovernmental organizations (NGOs), to name just the most active participants in global policy debates. In order to become influential, international secretariats need to actively compete for the attention of negotiators rather than trying to operate invisibly and underneath their radar. Moreover, due to negotiators' attention limits, international secretariats are more likely to attract the attention of national delegations if their problem definitions and policy preferences coincide with those brought forward by other organizations such as NGOs or scientific organizations.

Recent studies in the fields of international relations and international public administration have implicitly taken this attention-seeking character of international secretariats into account by focusing on their cognitive influence on international policy outputs (Biermann and Siebenhüner 2009). On the one hand, Depledge (2007) shows that treaty secretariats may provide policy-relevant information to negotiators by closely cooperating with the chairs or presidency of multilateral negotiations. On the other hand, Jinnah (2014) analyzes how treaty secretariats position themselves at the center of transnational communication flows that surround official multilateral negotiations, thereby providing policy-relevant information to negotiators from the outside (see also Jörgens, Kolleck, and Saerbeck 2016). In a similar vein, Abbott and colleagues (Abbott and Snidal 2010; Abbott et al. 2015) conceive of international public administrations (IPAs) as "orchestrators." Rather than trying to adopt and implement binding intergovernmental treaties, international organizations and their bureaucracies acting as orchestrators follow a complementary strategy of "reaching out to private actors and institutions, collaborating with them, and supporting and shaping their activities" in order to achieve their regulatory goals and purposes (Abbott and Snidal 2010: 315). Both approaches are similar to our notion of attention-seeking bureaucracies in that they expect IPAs to actively engage in issue-specific policy discourses within and beyond the intergovernmental decision-making that stands at the core of international organizations or multilateral negotiations.

However, studies of international organizations as orchestrators do not always draw a clear distinction between the broader international organization and the IPA as the permanent administrative body within it. In particular, they often fail to demonstrate that the outreach to private or subnational actors that characterizes orchestration is an autonomous initiative of the international bureaucracy and not mandated or encouraged by the international organization's member state governments. If international bureaucracies mostly act in line with their principals' preferences, that is, if their international organization's plenary or council back their efforts to orchestrate the individual actions of a wide range of transnational actors, then the distinction between international organization and IPA agency becomes blurred. By focusing on international treaty secretariats, that is, international bureaucracies that are not an integrative part of a broader international organization, we hope to be better able to explore the strategies that IPAs employ to provide policy-relevant information to decision-makers.

In the following, we describe two potential pathways through which international treaty secretariats may attract the attention of the official parties to multilateral negotiations, that is, supplying policy-relevant information to negotiators and building external support for their preferred policy outputs.

When looking at these two pathways of influence, one could easily be reminded of lobbying strategies that NGOs or business organizations might use to shape the political process according to their political goals. In some ways, these strategies may also resemble those of nation-states, who also build alliances with other authoritative actors in order to further their political goals. So what is the distinctively bureaucratic element of such attention-seeking behavior? In fact, there is an important distinction between the influencing strategies that state and nonstate stakeholders on the one hand may use and the attention-seeking strategy of IPAs on the other. IPAs employ this strategy based on their bureaucratic authority, which is the most important source of their influence. The bureaucratic authority IPAs enjoy sets them apart from other actors, since it helps their "voice be heard, recognized, and believed. This right to speak credibly is central to the way authority produces effects" (Barnett and Finnemore 2004: 20). Bureaucracies can be seen as the embodiment of rational-legal authority, which is a general and impersonal form of ruling that relies on legalities, procedures, and rules that offer order, classification, and a division of labor (Barnett and Finnemore 2004). Apart from this rational-legal foundation, IPAs furthermore enjoy legitimate authority due to parties' delegation of tasks to them, the shared norms or the "morality" that they defend, and their distinctive expertise, which can include an institutional memory concerning the treaty convention and technical and scientific, administrative and procedural, and normative and diplomatic knowledge (Barnett and Finnemore 2004; Bauer 2006; Busch and Liese 2017; Herold et al. 2021; Jinnah 2014; Littoz-Monnet 2017; Weber 2018; Wit et al. 2020). The effectiveness of bureaucratic authority based on these sources may further be enhanced by an IPA's display of leadership. Apart from the rational-legal authority of bureaucracies, Webererian social science points out the importance of charismatic leadership that is deliberately used to enhance a bureaucracy's authority and thereby leeway of action. The leadership component extends the concept of bureaucratic authority and adds a political element that goes beyond the mere technocratic role of a bureaucracy (Bauer 2006; Weber 2018; see also Section 2.3). We argue that attention-seeking treaty secretariats indeed make use of their bureaucratic authority understood as an entrepreneurial stance vis-á-vis their principals (Green 2014: 33; Well et al. 2020: 108).

## *Seeking Attention from the Inside: Treaty Secretariats' Cooperation with Chairpersons of Multilateral Negotiations*

The first pathway has been described in detail by Depledge (2007), who argues that treaty secretariats and chairpersons of multilateral negotiations are endowed with complementary resources, that is, political authority in the case of the chairperson and policy-relevant expertise as well as a certain distance to national governments and their domestically rooted preferences in the case of the secretariat. By combining their respective resources, secretariats and chairpersons can have considerable influence on the outcomes of multilateral negotiations. The secretariat assists the chairpersons in observing the lines of conflict that emerge between national delegations and propose compromises capable of overcoming policy divides and bringing negotiations to a successful end. Often this can be done through a reframing of the policy problem at stake or by bringing in new policy solutions that are more acceptable to reluctant negotiation parties than previously debated ones. Due to their expertise and their permanent monitoring activities, secretariats can provide valuable information to the chairs. Furthermore, due to their mandate as neutral and impartial actors, secretariats often refrain from claiming credit for their input. Chairs are free to use the input provided by secretariats in any way they intend. By taking on the ideas provided by the secretariat as their own, chairs endow them with the legitimacy needed to be heard by other negotiators.

Secretariats gain a privileged channel of communication to negotiators. By communicating with the chairs of convention bodies, who again directly address the negotiating parties, secretariats can significantly increase the probability that they are heard by negotiators, albeit in an indirect way. As Depledge (2007: 62) summarizes, "Chairpersons and secretariats are ... locked into a mutually interdependent relationship: the Chairperson often relies on the secretariat to provide the intellectual resources needed for him/her to exercise effective leadership, while the secretariat depends on an able Chairperson to provide the veil of legitimacy needed for it to input productively into the negotiation process." This symbiotic relationship does not mean that the negotiating parties are not aware of the secretariat's policy-shaping role in this process. In a large-scale survey we conducted in 2015 and 2016 among the participants of UNFCCC and CBD Conference of the Parties (COPs) (see also Chapter 9), we found that the two secretariats were trusted as providers of not only procedural information but also policy-related expertise.

Different variations of "supplying information from the inside" into the negotiation process are conceivable. These variations can be understood as subcategories of the internal pathway to gain influence described here. For example, when supplying policy-relevant information directly to chairs, a presidency, or parties, secretariats also contribute to *finding compromises* between opposing views. Moreover, it may be less important to supply additional information at a given time during or between negotiations than to *translate* the content of information into policy-relevant knowledge products, options for negotiation texts, or tactics. Information can be translated or applied to a political problem in such a way that it reflects the preferences of the secretariat. Such a translation activity goes beyond the pure passing on of information but can be as seen as shaping assumptions as bases for the actions of policymakers (Bijker and Latour 1988). Translation actions in the negotiation facilitation can therefore shape both the policy options and the policy discourses that give negotiations a certain character or direction. Another, similar, possibility is for secretariat staff to propose an *issue linkage*, that is, propose to look closely at a causal connection between one issue of the respective treaty, such as climate change or biodiversity, and an issue that is outside the realm of the treaty, such as health or security (Hall 2016: 6; Jinnah 2014: 67). Translation and issue linkage are forms of normative influence in that they can shape procedures, frame issues, and define participation (Biermann and Siebenhüner 2009). Finally, treaty secretariats may even go so far as to initiate the *production of information* they want to share with parties, for example, by commissioning certain studies.

### *Building Support from the Outside: Treaty Secretariats as Transnational Knowledge Brokers*

Multilateral environmental agreements are characterized by a multisectoral and a multiactor network structure. They can be described as "a system of continuous negotiation among nested governments at several territorial tiers – supranational, national, regional, and local – as the result of a broad process of institutional creation and decisional re-allocation" (Marks 1993: 392). They belong to the system of global environmental governance, which is marked by increasing complexity, polycentricity, and institutional fragmentation (Raustiala and Victor 2004; Zelli and van Asselt 2013). These dynamics are also driven by a proliferation of international institutions and treaties, all of which are managed by IPAs (Wit et al. 2020). Based on the phenomenon of multi-level reinforcement, which was first discussed with regard to the European Union (Schreurs and Tiberghien 2007), Ostrom (2010: 552) claims that the multilevel and multiactor systems of global climate governance propose important benefits in terms of fostering innovation, learning about policy alternatives, and achieving "more effective, equitable and sustainable outcomes at multiple scales." Thus, as Jänicke (2017) points out, it is a system which offers an opportunity structure in which skilled strategic action would allow an actor to mobilize support for ambitious policy objectives at different levels of governance and by a broad range of actors. One dimension of this opportunity structure is the emergence of governance voids, which can result in shifting actor constellations and rules of policymaking (Hajer 2003). Secretariats are well suited to fill such governance gaps, since their "unique position in governance networks ... allows them to operate in this political space" (Jinnah 2014: 48).

Attention-seeking treaty secretariats can strategically use this multilevel structure to help advance negotiations by acting as knowledge brokers that link broader transnational policy discourses to specific negotiation items. By linking actors that were disconnected before, this strategy may also lead to a form of issue linkage, that is, to a connection of a specific negotiation item with the broader policy concern of an external actor (Hall 2016: 6; Jinnah 2014: 67). The fact that IPAs can draw on their network position for their authority results from the diversified environmental governance architecture, where networks between organizations and actor types are increasingly important for effective governance (Jordan et al. 2015; Zelli 2018). Secretariat staff build up a dense web of relationships within and beyond their treaties and contribute to organizational learning (Kolleck et al. 2017; Varone, Ingold, and Fischer 2019). A similar role of bureaucracy has been observed at the national level by Fernandez and Gould (1994) in a study of the US health policy domain. This study finds that "occupants of ... 'brokerage positions' will be influential in policymaking to the degree that they facilitate communication among actors who would not otherwise interact" (Fernandez and Gould 1994: 1482). In a similar vein, Carpenter (2001) identifies organizational centrality, in this case defined as close ties with a large number of public and private organizations in a policy network, as one of the key factors that enable public administrations and hence treaty secretariats to play a brokerage role in issue-specific policy discourses. In a comparative study of three US federal bureaucracies, he shows that bureaucratic autonomy and influence increase with their centrality in broader issue-specific actor and communication networks. Providing linkages and knowledge sources (and even knowledge themselves), public authorities can act as intermediaries and hence knowledge brokers to promote issues and ensure cooperation in a specific issue discussed under a given framework (Christopoulos and Ingold 2015). In particular, in situations of pending stalemate in multilateral negotiations, secretariats can try to bring a new dynamic into the negotiation process by extending the policy debate to external actors who share the secretariat's general preference of a positive negotiation outcome as well as its specific problem perceptions and policy preferences. By deliberately extending issue-specific policy debates beyond the inner circle of official parties to multilateral negotiations (i.e., national delegations), we expect secretariats to try to build transnational support for the policy issues at stake, thereby raising pressure from the outside on national governments to continue and successfully conclude negotiations.[2] An important precondition for this second strategy is a strong embeddedness and centrality of international secretariats in the broader transnational policy networks that surround treaty negotiations. In the engagement with external actors for the purpose of attention-seeking, bureaucratic leadership particularly at the executive level becomes important. Biermann et al. (2009: 58) conceptualize "strong leadership" as the behavior of the leader of an international bureaucracy that follows a style of leadership that is "charismatic, visionary, and popular, as well as flexible and reflexive" (see also Chapter 5). Leaders' flexibility and openness to change and the ability to adapt their goals, international processes, and the organizational structure to perceived external challenges in learning processes are also considered to be essential for strong leadership in international bureaucracies (Biermann et al. 2009; Hall and Woods 2018).

In sum, we argue that convention secretariats are likely to employ a dual strategy to directly and indirectly draw the attention of negotiators to their own policy-specific knowledge and information. Convention secretariats may act either *directly and internally* via the chairpersons, presidents, or parties of multilateral negotiations or *indirectly and externally* via the broader transnational policy network that has evolved around the respective treaty. They may also opt for a combination of both strategies. The following case study of the activities of the CBD and the UNFCCC secretariats explores these potential pathways.

## 4.3 The Secretariats of the UNFCCC and of the CBD

In order to better understand the role of international treaty secretariats in issue-specific multilateral negotiations, how they interact with and whether they attract the attention of member states (parties to the convention) and nonparty stakeholders,

---

[2] The underlying logic of this strategy is similar to what Keck and Sikkink (1999: 93) in their work on transnational advocacy networks describe as the "boomerang pattern of influence," that is, a strategy where "NGOs may directly seek international allies to try to bring pressure on their states from outside."

this section follows an inductive and exploratory approach. Methodologically, we drew on twenty-one qualitative semistructured expert interviews with staff of the UNFCCC and CBD Secretariats from different hierarchical levels and analyzed documents of UNFCCC and CBD negotiations using qualitative content analysis. Furthermore, we drew on our participant observations of these negotiations between 2014 and 2022. Interviews with the UNFCCC Secretariat are marked "1–7A" and those with the CBD Secretariat "1–14B" throughout the analysis. Relevant documents include statements issued by the secretariats, party submissions, published papers, and interviews related to the multilateral treaty conferences. These documents were analyzed as representative material of what the secretariat supports to be its key message and mode of interaction with other actors. Semistructured interviews were chosen as an adequate tool for conducting expert interviews, since they can detect both specific and context-related knowledge and thereby address both the practical and discursive consciousness of the interviewees (Meuser and Nagel 2009). Specific knowledge relates to an expert's own actions concerning the policy process in the CBD and the UNFCCC, while context-related knowledge refers to the actions of others, such as stakeholders active in the wider context of the CBD and the UNFCCC. Interviewees were queried, among others, about the role and activities of the secretariat during and between negotiations as well as their relationship to the respective chairpersons, party delegates, and nonparty stakeholders and their motivation for being engaged in the multilateral negotiations.[3] Since interviewees naturally report their own perceptions of events, validating these with participant observations and document analysis was an important additional step (Creswell 2009). The interviews were transcribed, anonymized, and combined with the collected documents. The qualitative data gathered from the documents and interviews was analyzed using inductive techniques of qualitative content analysis following Mayring and Frenzel (2014). The process of coding followed the rules of qualitative content analysis. Codes were related to the way the international treaty secretariats report to interact with other stakeholders and to shape the global agenda concerning the CBD and the UNFCCC.

The following section analyzes the biodiversity and the climate secretariats' roles within the multilateral negotiations and their use of interaction strategies. Firstly, we find direct attention-seeking strategies, which rely on the internal cooperation between the secretariats and the chairpersons, COP presidency, or party delegates. Secondly, we find indirect attention-seeking strategies, which secretariats employ by engaging with a wide range of actors in the broader transnational policy debates surrounding the formal climate and biodiversity negotiations.

---

[3] The analysis of expert interviews focuses on thematic units, meaning text extracts with similar topics, which are scattered over the interviews. The comparability of the interviews is ensured by the commonly shared context of the experts, as well as by the interview guidelines (Meuser and Nagel 2009: 35).

## Direct Attention-Seeking within Multilateral Negotiations

### UNFCCC

The climate secretariat originally has a very specific and rather technocratic mandate to support the UNFCCC negotiations, which are "party-driven" (A1–A5, A7; UN 1992b). Climate negotiations tend to be contentious and have in the past at certain times been on the verge of collapsing, while at the same time being under the pressure of delivering an ambitious result considering the potential for irreversible and catastrophic change (Depledge 2005: 20; Kinley 2017). Given this situation – highly politicized, stalling negotiations in the context of high political expectations to deliver an ambitious result – the climate secretariat has in the past drawn attention to its ability to perform tasks that go beyond its classical role of acting "like a secretary" in the background (1A, 6A, 7A; Well et al. 2020). In 2021, former executive secretaries and senior staff of the climate secretariat published a journal article entitled "Beyond Good Intentions, to Urgent Action: Former UNFCCC Leaders Take Stock of Thirty Years of International Climate Change Negotiations." One of their key messages aims to drive the attention of policymakers toward what they, according to their experience as former executive staff, deem necessary: "'Business as usual' in climate change negotiations will mean failure to avoid dangerous climate change. Fuller engagement by leaders is crucial to ensuring an all-of-government approach. The UNFCCC process should address its unwieldiness and act in line with the urgency of the issue" (Kinley et al. 2021: 593). Although this was published by a group of former executive secretaries, it is in line with the increasingly vocal and attention-seeking role the climate secretariat assumes.

This section will sketch the evolution of the climate secretariat's attention-seeking behavior in the context of the negotiations leading up to the Paris Agreement in 2015 and during the "post-Paris" years. In this section we aim to strengthen our argument that the climate secretariat not only is the organizational backbone to the negotiation process but increasingly draws attention to its problem-solving strategies and substantive preferences, thereby contributing to agenda-setting, policy-drafting, and reaching consensus among states. Such actions can be directed to the conference presidency, chairpersons, or delegates directly.

### Crafting the Paris Agreement

When trying to explain what enabled the negotiation of the Paris Agreement at COP21 in 2015, studies point to factors such as civil society mobilization (Jacobs 2016), great power politics (Milkoreit 2019), leadership (Eckersley 2020), and institutional design (Allan et al. 2021) but also to the careful management and the "diplomatic process and entrepreneurial leadership by host governments" as well as to their "timing, pacing, sequencing and coordination of sessions, as well as the strategic rhetoric" (Dimitrov 2016: 9). While these actors and factors have been credited for the successful negotiations, it is worthwhile to also take into the account the contribution of the climate secretariat, despite its technocratic mandate. Allan et al. (2021: 25) identify certain entrepreneurial actors that were crucial for finalizing the Paris Agreement. Apart from the role of the COP presidency and states with political clout, they point to the entrepreneurial role played by the secretariat:

> The strategies of specific actors in the negotiations ... proved crucial to securing the final components of the deal: the 1.5°C target and the ratchet-up mechanism. These were key demands of vulnerable countries, and crucial for agreement. Without their sign-on, a Copenhagen-level fiasco may have occurred. However, others played an important role in steering parties toward common ground. Here, therefore, we highlight the entrepreneurship of several actors for the overall design: the French COP Presidency and the UNFCCC Secretariat, US and Chinese diplomats, and those in the High-Ambition Coalition.
> 
> *(Allan et al. 2021: 15)*

This entrepreneurial role of the climate secretariat is also corroborated by interviews with secretariat staff. One member of the secretariat's staff describes its role during negotiations by way of comparison: "The UNFCCC is very different from other processes. If you look at the Security Council, it is the Parties who bring the text and ... negotiate around that. ... In the Climate Change Convention, ... the secretariat plays a big role ... [in] preparing all the drafts" (1A). Relying on their expertise and experience, the climate secretariat acts as an intermediary between parties' interests on the one hand and the chairs' and presidency's organizational tasks, which include compiling and presenting a draft decision text reflecting these positions on the other hand (1A, 3A, 7A). To this end, secretariat staff seek their attention by offering procedural advice as well as substantive information and highlight possible areas of compromise or "landing zones," that is, the likeliest compromise on core issues, all of which help parties when drafting decision texts (see also Allan et al. 2021: 16). Secretariat staff were able to form trustful personal relations and to gain the attention of delegates, as one member of staff recalls: "Because of the personal relationships that were built during the process, at this working level you stop seeing people as the guy from France, the guy from Brazil, but we are just the guys that are trying to ... draft a text. ... I would sit with the people, not with the countries" (1A; similarly 3A, 7A). Such personal relations also enable the secretariat to foster the trust of parties into the UN multilateral process: "Trust breaks down for many reasons. We try to bring people together, if governments walk out of a session because of loss of trust in the process or each other. Usually, the secretariat tries to meet with them, ... and create a frame where people talk to each other again" (4A).

What is more, in cases of technical or highly politicized issue areas, such as climate change mitigation, the negotiations may be "so complicated that chairs do not have any other option but to go along with the drafts they receive" by the secretariat (1A). Usually, such a secretariat-prepared text would be tabled by the chair, thereby combining the secretariat's policy-relevant expertise with the chair's political authority, who can together gain considerable influence on how negotiations develop. However, the following example shows that the climate secretariat is able to play this role on its own. The negotiations leading to the Paris Agreement combined low levels of trust between negotiation parties and a high degree of politicization and technicality of the agenda items, leading to long and barely readable draft decisions, containing multiple unresolved issues and options (1A, 7A; Dimitrov 2016). In this situation, "the visions were so stark, that you didn't have a possibility to work on a text tabled by any party" and the "trust was so bad, that not even the chairs were asked to do it" (1A; see also Allan et al. 2021: 16). When referring to a section of the text that was later included into the Paris Agreement, this staff member reports that "[t]he decision was entirely drafted by us" (1A). This account shows that the secretariat was able to directly contribute to the final text of the Paris Agreement, having drawn attention to its relevant expertise and earned the trust of parties to assist in this way beforehand.

While this may not be the usual course of how negotiations are organized as it exceeds the designated role that the climate secretariat has in multilateral negotiations, this example does show that circumstances such as high politicization and technicality and low trust between states have been conducive for the climate secretariat as an attention-seeking bureaucracy. It gained the attention of chairpersons, the conference presidency, and negotiation parties by reducing the complexity of technical negotiations, synthesizing positions, and offering a line of compromise. It was then possible to feed procedural advice, substantive information, and even draft text into the process. Such an attention-seeking behavior enabled the climate secretariat not only to contribute to the successful completion of negotiations but also to leave a fingerprint on the outcome of the final text, as in the case of the Paris Agreement.

### Supporting the Post-Paris Architecture

While this type of direct attention-seeking before and after COP21 could be observed by means of participant observation and expert interviews, it was a behavior that stayed within the confines of the relationship between parties and the secretariat and was not openly displayed beyond this professional environment. However, since 2017, the secretariat has published annual reports, in which it reflects on its changing role vis-à-vis parties and nonparty stakeholders, which is marked by a focus on implementation and a stance that acknowledges a more visible role for itself: "While the secretariat in its early years focused on facilitating intergovernmental climate negotiations, today it supports a complex architecture that serves to advance the implementation of the Convention, the Kyoto Protocol and the Paris Agreement" (UNFCCC 2020: 8). In the currently (as of January 2022) available reports of 2017 to 2020, it reports on its own activities during the year in relation to important negotiation achievements as well as its support for implementation and capacity-building.[4] It also sheds light on how it supports parties through translation of information into policy-relevant advice, by proposing or supporting issue linkages and by providing guidance to parties. For example, in its 2019 annual report, the secretariat reports to have "launched efforts to help Parties prepare to implement the enhanced transparency framework" (UNFCCC 2020: 15) established under the Paris Agreement, which provides guidance to countries on how to report progress on their climate change mitigation, adaptation, and relevant support to or from other countries. The support by the secretariat included providing technical support on the implementation of the enhanced transparency framework, designing institutional arrangements to support it, providing guidance on nationally determined contributions, and producing detailed expert training materials on national greenhouse gas inventories (UNFCCC 2020). This support potentially has a far-reaching impact on how parties implement the enhanced transparency framework, since it helps to turn the relevant provisions in Article 13 of the Paris Agreement into national policy tools. The secretariat openly acknowledges this: "The secretariat plays a crucial role in putting into practice the transparency and accountability arrangements for climate change reporting" (UNFCCC 2020: 8). Similarly, the secretariat reports to support parties on a wide range of processes related to adaptation, stepping in when needed: "[I]n the face of decreasing financial resources, the secretariat facilitated the [Adaptation] Committee's communication and outreach activities" (UNFCCC 2020: 17).

While this emphasis on implementation and capacity-building is one important dimension of the role of the climate secretariat since the Paris Agreement has come into effect, a second important development is issue linkage between climate change and other policy areas. As explained earlier, issue linkage can be an element of direct attention-seeking and normative influence. Jörgens, Kolleck, and Saerbeck (2016) described the role of the climate secretariat for supporting the link between gender and climate change. A more recent example of issue linkage is the secretariat activities in the area of climate and security. Since 2007, states have increasingly discussed the link between climate and security at the United Nations Security Council (Abdenur 2021). Although it has not been an agenda item or prominent angle in the context of UNFCCC negotiations, discussions on it

---

[4] These can be found at https://unfccc.int/annualreport

have increased recently during official side events, pointing out the different security implications of climate change, such as risks for social stability (e.g., Climate Diplomacy 2018). At COP25 in Madrid, the climate secretariat hosted a side event entitled "Dialogue on climate-related risks to social stability: law and governance approaches" (UNFCCC 2019; participant observation at COP25). By hosting this as a secretariat-sponsored event and providing a framing on climate and security "from the inside," the secretariat drove the attention of delegates to the link of climate and social stability and provided support to considering the effects of climate change from this perspective. It invited the chair of the Subsidiary Body for Scientific and Technological Advice as well as actors who favor the angle of climate-related risks to social stability, such as the United Nations Convention to Combat Desertification Secretariat, the Office of the High Commissioner on Human Rights Secretariat, and representatives of Ghana and Germany (both founding members of the Group of Friends on Climate and Security in the Security Council) (Federal Foreign Office 2018; participant observation at COP25). This is an example of the climate secretariat's open support for the link between climate and security, which is still not an agenda item under the UNFCCC and therefore not mandated, but is certainly in line with highlighting the "planetary emergency" that climate change poses (see, e.g., UNFCCC 2020: 6).

Summarizing, we observe that the direct attention-seeking behavior could be observed in the run-up to the Paris Agreement and has since become more pronounced, public, and part of a broader communication and engagement strategy, blending into the indirect attention-seeking of all stakeholders. This will be dealt with in depth in the next subsection.

### CBD

The biodiversity secretariat seeks the attention of parties directly throughout the whole policy cycle: It contributes to agenda-setting by alerting parties to new policy issues or possible linkages; it provides input into the negotiation process by seeking attention for its analysis of lines of compromise during policy-drafting; and it supports parties in the implementation of decisions by providing capacity-building. The following section will lay out how the interviews substantiate these findings.

In the case of ocean governance, for example, the biodiversity secretariat actively seeks the attention of parties in order to put the issue on the agenda and create a mandate for its own activities through COP decisions. For example, when certain parties showed interest in aspects of ocean governance, such as ocean acidification and marine mining, the secretariat responded to this initial interest by trying "to make it an issue" at a larger scale. Secretariat staff tried to "find a way for an issue to gain attraction at policy level, and ... find an excuse to help a country ... so that the issue rises, and finally the COP will reapprove the importance and maybe even request the secretariat to do more" (10B, 11B). The role of the secretariat in this strategy is to highlight the global implications and benefits of specific topics, such as the role of a healthy ocean for many dimensions of sustainable development, as well as to "see issues in perspective, to connect relevant partners." If this strategy of translation and agenda-setting is successful, the secretariat may have created an own role for the issue in question: "Once they are in, we try to serve them" (10B, similarly 3B). Secretariat staff also reported helping parties and nonstate actors in framing ocean-related topics, in order to create a fit with national debates and contexts, thereby also promoting certain frames, such as looking at ocean areas from different continents as a whole. One staff member formulated this approach as "Forget your box and see the environment as a whole" (10B).

While the climate secretariat cannot attract the attention of specific parties, for example, by organizing workshops that target only one or few parties, the biodiversity secretariat can organize national workshops on specific issues if parties express a special concern for these topics, such as for the issue of marine mining. Sensitive to the worries of specific parties, secretariat staff assisted with the provision of an impact assessment and the invitation of experts and stakeholders for this issue, thereby drawing attention to its expertise, network, and convening power. According to several interviewees, such activities can pave the way for outputs that help to advance the negotiations, such as the compilation of national long-term visions for all stakeholders (1B, 6B, 7B, 10B). In this sense, the biodiversity secretariat can benefit from a wider mandate than the climate secretariat to attract the attention of specific parties and support them according to their needs. We will describe the biodiversity secretariat's mandate in more detail here.

In terms of policy-drafting and cooperation with chairpersons, the biodiversity secretariat is similar to the climate secretariat. It is also tasked with providing logistical and procedural support in negotiations (Art. 24 of CBD). Nevertheless, it actively contributes to negotiations by pointing out the benefits of mutual cooperation, suggests substantive or procedural solutions to negotiation deadlock, and shows parties what they would miss out on or maybe even lose control over if they do not cooperate (1B, 3B, 6B, 10B). To reach an agreement in negotiations, the secretariat "create[d] a fear of being left out" (10B) until parties decided to cooperate. One member of staff reported attracting especially the attention of those parties that occupy veto positions or otherwise block progress in negotiations: "The most difficult they are, the most helpful I am," following the credo that "going backwards is no option" (3B).

Seeking the attention of chairpersons was also key, for example by providing a "choreography" of meetings, which included not only background information on the positions of delegations and potential pitfalls concerning specific agenda items but also suggestions on how to navigate such pitfalls and opposing interests (1B, 9B, 5B). By providing such procedural advice, the biodiversity secretariat actively sought to feed its own policy preferences into the negotiations and build compromise. A member of staff would not "go [into negotiations] with a blank page, but make[s] suggestions how to frame, how to make it work" (3B). In particular, if agreement among negotiators is hard to achieve, the secretariat "give[s] parties options what they could agree on" (3B). "You incorporate ... as much as you can" (1B) while ensuring that the suggested policy options "reflect a balance of [voiced] views" (1B, 3B, 5B).

While the biodiversity secretariat has no mandate for implementation, it is able to assist and support parties in implementing decisions and working on their National Biodiversity Strategies and Action Plans by providing capacity development: "I think we can say without hesitation that the countries do get a lot of help from the CBD staff" (2B, similarly 3B, 7B, 8B, 10B). Especially parties from least developed countries, small island countries, and indigenous and local communities are supported frequently with the goal of empowering them to effectively play their role in the negotiation and implementation process: "We need to build everyone's capacity at all levels" (3B). Its role in capacity development and in assisting the implementation of decisions is a further avenue for the biodiversity secretariat to seek attention for its expertise and policy suggestions.

### *Indirect Attention-Seeking via the Policy Network*

#### UNFCCC

Directly seeking the attention of parties to the UNFCCC is viable for the climate secretariat with regard to concrete negotiation topics and processes. It does so by adopting a strong role in policy-drafting, organizing negotiation sessions, and building trust, as pointed out earlier. However, when wishing to attract the attention of parties regarding broader perspectives on combatting and adapting to climate change, such as connecting climate change to economic and societal questions, the climate secretariat attracts the attention of parties in an indirect way, by conveying its messages through the extensive transnational policy network that has evolved around the UNFCCC. The climate secretariat holds a central position in the relevant issue-specific information flows and transnational cooperation networks, enabling it to act as a broker of information between actors outside the formal negotiations, such as NGOs, think tanks, research institutions, private sector organizations, international organizations, and the parties themselves (Saerbeck et al. 2020). Using this central network position, the climate secretariat can provide substantive and procedural information to well-connected stakeholders, resulting in an excellent reach of its messaging (Saerbeck et al. 2020; 1A, 3A, 4A, 6A). By gathering, synthesizing, processing, and disseminating policy-relevant information that went beyond the negotiation of specific decision drafts to a wide range of different stakeholders, the climate secretariat attempted to connect broader policy discourses with specific negotiation items.

*Giving a Sense of Direction in the Run-Up to the Paris Agreement*

Using this network position, the secretariat aimed to change the "narrative" of how climate action could and should be viewed (6A) prior to COP21. Staff members wanted to demonstrate that the negotiation process "was part of a bigger transformation going on" (6A). The secretariat aimed to streamline the policy discourse, to make it more coherent and forward-looking, because "people weren't really getting it, ordinary citizens, many governments, particularly the negotiators … were all running in different directions," as one senior member of staff remembers, adding, "have you ever seen the Monty Python video of the Olympics for people that have no sense of direction, then you know exactly what I am talking about" (6A). It provided orientation for example by directing attention to successful climate policies already in place before COP21. Giving such a "sense of direction" was the goal of a communication strategy that aimed at attracting the attention of parties indirectly by targeting prominent and well-connected societal and political actors. The positive message of this communication strategy was introduced into the "political landscape of the year," including G20 and G7 meetings, World Bank and the International Monetary Fund (IMF) meetings, and even meetings of religious groups in order to mainstream this message into different policy fields (6A; G7 Germany 2015; G20 Australia 2014; Lagarde 2014; Mou 2015; World Bank 2015). To this end, the secretariat partnered with important stakeholders and public figures or organizations for them to "carry" and "amplify [the] message" of "how well cities are doing on climate change, … how big corporations like Unilever are greening their supply chains," to name two examples (6A).

In line with this strategy, the executive secretary incumbent from 2010 to 2016, Christiana Figueres, sought the attention of parties by starting her climate diplomacy campaign ahead of the negotiations of COP21. One indirect way to do this was by thanking cities, faith groups, companies, investors, and other nonparty stakeholders publicly for going ahead with innovative climate activities while at the same time asking for more ambitious actions (6A). Another one was to ask prominent individuals to speak out about climate action, including a meeting with the Pope to discuss how climate change could figure prominently in his encyclical "Laudato si'" (6A; King 2014). She reached a multitude of actors and also addressed parties "through her social media account, she would thank India for saying they would invest in solar. She would thank … Johannesburg, for committing to a certain target on climate change," thereby drawing attention to "all the benefits that come with climate change [policies], all the positive outcomes that can come by a low-carbon transition" (6A). Questions that were not officially on the negotiation table but that were nonetheless crucial in achieving emission reductions could be included into the policy debate (3A, 2A, 4A, 5A, 6A). For example, "Momentum for Change" was initiated by the climate secretariat in 2013 to connect different economic and societal sectors to climate change action by publishing information on "lighthouse activities" of climate action and low-carbon development and by awarding the UN Global Climate Action Awards annually (UNFCCC, 2014; see also Chapter 3). A recent strand of literature describes initiatives by the climate secretariat to include nonparty stakeholders, such as Momentum for Change, the Non-state Climate Action Zone for Climate Action, the Lima–Paris Action Agenda, the Marrakech Partnership for Global Climate Action, or Action for Climate Empowerment, as orchestration (Hale 2016; Thew, Middlemiss, and Paavola 2021; see also Chapter 3).

The goal of such an indirect attention-seeking behavior via the transnational policy network was twofold: First, ideas and information were distributed through an additional, powerful channel, thereby building transnational support for climate action and raising pressure on national governments to agree on ambitious climate policies from the outside. Second, through this informal channel that was independent of narrowly phrased agenda items and a legalistic negotiation logic, fresh ideas could be circulated. Looking back at COP21, one former senior official of the climate secretariat noted in 2016 that "policy announcements and initiatives made outside of the formal negotiations were also spectacular in scale and scope, suggesting that a new sustainable growth model is underway" and that nonstate actors in the Paris Agreement "are increasingly becoming the engine of both mitigation and adaptation action. This is helping to define a 'new normal'" (Kinley 2017: 4). Through its strategy of engaging and empowering nonparty stakeholders and conveying its own policy preferences through this network (2A, 3A, 4A, 6A), the climate secretariat has arguably contributed to the necessary "cognitive change" that enabled the Paris Agreement (Dimitrov 2016: 1). It ensured that those "persuasive arguments about the economic benefits of climate action" that "altered preferences in favor of policy commitments at both national and international levels" (Dimitrov 2016: 1) found their way into the policy debate and onto the agenda.

*Executive Leadership and Legitimacy Concerns*

The extent of indirect attention-seeking and influence-seeking behavior of the climate secretariat varies over time and according to the political context of global climate governance. In 2009, Bauer, Busch, and Siebenhüner found the autonomy and influence-seeking behavior of the climate secretariat to be extremely limited, if existing at all: "That staff at all levels have internalized the expectations of parties and the resulting lack of leadership further explains the limitation of its influence. In fact, the secretariat has accepted the parties' definitions of boundaries and 'has very rarely attempted to exercise open substantive leadership by brokering agreements among parties'" (Bauer, Busch and Siebenhüner 2009: 179). This description stands in stark contrast to the leadership displayed by the executive secretary in particular before COP21. Figueres (2013: 538) highlighted in an article: "The only way to regain energy security, stabilize water and food availability, and avoid the worst effects of climate change is to accelerate the economic tipping point towards low-carbon growth, towards the point where low-carbon living is the norm and not the novelty," thereby sketching her vision of how national climate policies should be spelled out. Thinking back to her first press conference in 2009, she reflects on how it was possible to achieve a global climate change agreement in an interview in 2016: "Impossible is not a fact, it's an attitude. … And I decided right then and there that I was going to change my attitude and I was going to help the world change its attitude on climate change" (Greene 2016). These statements show the departure from an attention-avoiding and neutral stance toward an attention-seeking and outspoken behavior, by which the secretariat deliberately stretched and surpassed the parties' definition of boundaries. In addition, Figueres' ability to adapt the goals and organizational processes of the UNFCCC secretariat to the challenges she identified and her aptitude in translating this into an effective strategy for engaging with a wide network of different actors made her leadership flexible, reflexive, and visionary. This kind of executive bureaucratic leadership was an important element of the attention-seeking activities of the secretariat especially vis-à-vis external actors in the run-up to and follow-up of the Paris Agreement.

Until today, we can observe different examples and varying degrees of attention-seeking behavior of the climate secretariat. While tracing this development in detail lies outside the scope of this empirical section, it is plausible that the initial attention-seeking behavior originated in the "fiasco-like" COP15 in 2009, which was "perceived to be constrained by the lumbering UNFCCC process that was limiting, rather than enabling climate action in a timely and responsive manner" (Dubash and Rajamani 2010; see also Figueres 2013). This "hurt the legitimacy of the UNFCCC" (Allan et al. 2021: 19) and the trust into the climate secretariat was lower than before COP15 (4A; Sommerer et al. 2022: 95, 177). As typical for a bureaucracy, it is likely that the climate secretariat sought the attention of parties and nonparty stakeholders also for the stake of self-preservation, by drawing attention to itself as an actor legitimized by visible policy outputs, for example, by assuming the role of an orchestrator with regard to nonstate climate action (Sommerer et al. 2022: 177).

This section has shown that, so far, the culmination of the climate secretariat's indirect attention-seeking behavior is the described effort leading to the Paris

Agreement. Since the adoption of the Paris Agreement the secretariat has continuously sought the attention of citizens and policymakers (Mederake et al. 2021; Saerbeck et al. 2020) and invested into a targeted communication strategy, increasingly online and via social media channels (UNFCCC 2020). Engaging with youth stakeholders represented by prominent persons such as Greta Thunberg fitted especially well into the strategy of including nonparty stakeholders as an integral pillar of the post-2015 climate regime (Thew, Middlemiss, and Paavola 2021). Instead of acting invisibly or from behind the scenes, part of the "new normal" of international climate administration is the climate secretariat's aim to garner trust into its work by indirectly seeking the attention of parties and nonparty stakeholders through its policy network.

## CBD Issue Linkages: Connecting with Relevant Policies

Since the biodiversity secretariat has the mandate to play a coordinating role, or that of an "overlap manager" in the biodiversity regime (Jinnah 2014: 73), seeking the attention of policymakers via both the intergovernmental and the transnational policy network, that is, via other international organizations and nongovernmental stakeholders, is a natural option for the biodiversity secretariat. The objectives of the CBD are biodiversity conservation, sustainable use of its components, and equitable sharing of its benefits (UN 1992a: Art. 1). These objectives overlap with a multitude of other multilateral environmental agreements that form the global biodiversity regime (Jinnah 2014: 68; Raustiala and Victor 2004: 277). With regard to engaging with other international bodies, the biodiversity secretariat has the mandate to actively seek the attention of international entities that overlap with these objectives (Jinnah 2014: 73). The CBD convention text states that the secretariat's functions shall be, inter alia, "to coordinate with other relevant international bodies and, in particular to enter into such administrative and contractual arrangements as may be required for the effective discharge of its functions" (UN 1992a: Art. 24[d]). It furthermore asks of parties to "contact, through the Secretariat, the executive bodies of conventions dealing with matters covered by this Convention with a view to establishing appropriate forms of cooperation with them" (UN 1992a: Art. 23, 4[h]).

Our analysis shows that the CBD Secretariat seeks attention in the transnational policy debates on biodiversity to increase the general weight of its arguments, build issue-specific coalitions with other stakeholders, and, in the long run, shape parties' preferences on substantive issues, including by issue linkage (1B, 3B, 6B, 8B, 13B). This includes liaising with international organizations on overlapping issues and linking the respective biodiversity issue to those of the broader policy concerns of other organizations. Such overlapping issues between the CBD and the UNFCCC are especially relevant, for example, forests, oceans, blue carbon (i.e., carbon stored in marine ecosystems), gender equality, and geoengineering

(1B, 9B, 10B, 13B, 14B; van Asselt 2011). Also, in the case of the causal relationship between climate change and biodiversity itself, the biodiversity secretariat deployed "an aggressive marketing campaign," in order to draw parties' attention to biodiversity conservation as a climate adaptation strategy (Jinnah 2014: 94; see also 13B). The UNFCCC has recently put an emphasis on "nature-based solutions," which reflects the link between the two conventions and recognizes "the interlinked global crises of climate change and biodiversity loss" and "the importance of ensuring the integrity of all ecosystems, including forests, the ocean and the cryosphere, and the protection of biodiversity" (UNFCCC 2021).

Other international organizations and, by extension, policy communities the biodiversity secretariat collaborates with include the United Nations Educational, Scientific and Cultural Organization, the Food and Agriculture Organization, the United Nations Environment Program, the World Conservation Monitoring Centre, and the World Meteorological Organization (7B, 8B, 9B, 10B, 13B, 14B). In order to liaise with the two other Rio Conventions, the United Nations Convention to Combat Desertification and the UNFCCC, the biodiversity secretariat is very active in the so-called Joint Liaison Group (13B). This is an institutionalized mechanism through which the executive heads and other members of staff of the three Rio Conventions meet to discuss and draw attention to overlapping issues between them (SCBD 2006). The CBD is furthermore deeply intertwined with the development, agricultural, and trade regimes, which are some of the most responsible sectors for biodiversity loss, as well as with the 2030 Agenda for Sustainable Development (Miller Smallwood et al. 2022: 48–49). Reaching out to organizations in these adjacent but also nonenvironmental policy fields provided the biodiversity secretariat with ample opportunity to link biodiversity to different issues and bring these connections to the attention of state actors. Framing biodiversity issues in the light of a connection to a different policy field may also attract the attention of actors outside of the biodiversity community and thereby inform and influence the public discourse. For example, the COVID-19 pandemic dramatically brought the connection between biodiversity and human health into focus, as the incumbent executive secretary Elizabeth Maruma Mrema highlighted in her opening statement for COP15 in 2021: "Now more than ever, we are witnessing a deep shift of awareness of the interconnected biodiversity, climate and health emergencies that we face. The COVID-19 pandemic is a stark reminder of the connection between human health, the health of species and our ecosystems." (SCBD 2021a)

An important avenue of reaching biodiversity goals is to mainstream them into other sectors and nonenvironmental policies, for example, by linking biodiversity and business practices (1B, 12B; SCBD 2016). Building on the interest of parties, the biodiversity secretariat launched several business-related events from 2005 on, which have become more numerous and prominent in recent years and "acted as a catalyst for larger discussions on business engagement issues and COP business decisions" (SCBD 2022a), such as the *Business and the 2010 Biodiversity Challenge*, the *Business and Biodiversity Forum*, the *Global Partnership for Business and Biodiversity*, and the *Business and Biodiversity Week* in 2021 (12B; Hickmann and Elsässer 2020; SCBD 2022a). Through coordinating and collaborating with companies, business associations, and civil society actors, the secretariat indirectly sought the attention of parties to bring the linkage between biodiversity and business into the spotlight (12B). Parties became gradually more interested and asked the secretariat at COP10 to establish a forum for them to interact with businesses and other stakeholders, which led the secretariat to launch the *Global Partnership on Business and Biodiversity* (SCBD 2010). In further decisions, the COP asked the secretariat to expand this work, including by liaising with other relevant organizations and by providing relevant capacity-building, tools, and guidance (SCBD 2021b) These activities are now listed under the umbrella of the *Business Engagement Programme* run by the secretariat and funded by the European Union, thereby further formalizing this issue linkage (SCBD 2022b).

### Nonstate Actor Engagement: Broadening the Discourse

The CBD furthermore reaches out to an array of nonstate actors, in order to support their participation in the policy process and create support for ambitious negotiation outcomes from the outside (1B, 3B, 6B, 8B, 10B, 11B, 12B, 14B). The CBD has a long history of engagement with stakeholders and stands out in this respect compared to other organizations in global environmental governance (Miller Smallwood et al. 2022). Nonstate actors are often more supportive of ambitious biodiversity policies than national delegations and can be key partners for implementation and accountability in the CBD (10B; see also Miller Smallwood et al. 2022: 57; Ulloa 2022). Therefore, the biodiversity secretariat builds transnational support for biodiversity topics by opening debates on certain agenda items to include broader concerns represented by civil society. Particular emphasis is placed on the cooperation with indigenous peoples and local communities (IPLCs), which may be viewed as "elders of the convention" (3B), which speaks to their sincere commitment to biodiversity conservation, excellent organization and knowledge of the negotiation process, dedication to cooperation, and, in many cases, low turnover rates (as opposed to national delegates, who have higher turnover rates) (3B, 10B). Target 18 of the 2020 Aichi Biodiversity Targets states that by 2020 traditional knowledge, innovations, and practices are to be respected and protected, and fully integrated and reflected in the implementation of the CBD (SCBD 2010). This makes IPLCs a key grouping of stakeholders through which the secretariat can advocate for an ambitious outcome of negotiations (1B, 3B, 10B). The CBD Secretariat also strives to empower regional actors, religious groups, research institutions, and universities to effectively participate in negotiations and other CBD events (3B, 8B, 10B). As described in the previous section, secretarial outreach activities furthermore include the private sector.

Such a strategic use of its embeddedness in broader policy discourses is in line with the findings of other studies that point out IPAs' potential roles as knowledge brokers or orchestrators (Abbott et al. 2015). Our findings add on to this since we see a particular emphasis on their agenda-setting role in instances of multilateral policy formulation. Our explorative study indicates that the secretariat of the CBD seeks the attention of a wide range of stakeholders outside of the convention on specific issues discussed under the framework of the CBD. It is the hub of a widespread stakeholder network, allowing secretarial staff to act as a knowledge brokers and enabling it to drive negotiations forward from the outside (see also Hickmann and Elsässer 2020; Mederake et al. 2021). In its increasing integration of nonstate actors into the CBD process, the secretariat follows a broader trend in global environmental and sustainability governance of collaborating with transnational actors (Kok and Ludwig 2022; Pattberg, Widerberg, and Kok 2019).

## 4.4 Conclusion

In this chapter we developed the contours of a heuristic framework for modeling the role and social interactions of international treaty secretariats with regard to issue-specific negotiations of multilateral treaty conferences. We drew on an explorative empirical study to illustrate the plausibility of our model. Overall, the empirical observations are in line with the theoretical framework outlined in the beginning. They show that international secretariats regularly act according to a logic of attention-seeking. Rather than withholding policy-relevant information from their principals or forming covert alliances with selected states, they act openly with the aim of increasing policymakers' awareness of their problem definitions and policy proposals. Seeking the attention of policymakers directly and internally as well as indirectly and externally proves to be a potent strategy of progress in the climate and the biodiversity regimes, confirming that bureaucratic behavior can alter knowledge and belief systems, thereby enabling political change (Barnett and Finnemore 2004). Attention-seeking international bureaucracies contribute to blurring the line between international politics and bureaucracy. Both the climate and the biodiversity secretariats successfully compete with other organizations, indeed with a whole industry of knowledge providers, in the provision of policy-relevant information to national bureaucracies and their political leadership. Among these organizations are other international organizations that are mandated to work on related issue areas as well as an array of actors from civil society and the private sector. And unlike other actors in global environmental governance, they can use their bureaucratic authority to this end. Both secretariats act as agenda-setters, policy entrepreneurs, and policy brokers, thereby furthering and shaping the negotiations in the respective conventions and including actors outside of the conventions into the policy debate. The climate secretariat exploits its narrow mandate by seeking attention for its policy solutions in negotiations and by rallying support for climate action in the transnational network, for which its central network position is key. The biodiversity secretariat has a slightly more lenient mandate and can also form alliances with individual or groups of parties and stakeholders. With a strong role in capacity development, it is also able to leave a mark on the policy implementation phase, albeit indirectly.

Our findings are also in line with empirical studies on the autonomy and influence of bureaucracies at the domestic level of the United States (Carpenter 2001; Workman 2015). We therefore argue that conceptualizing public administrations as attention-seeking actors can provide a fruitful complement to theories of delegation and oversight when studying the autonomy and influence of domestic bureaucracies.

Analyzing the role of bureaucracies at earlier stages of the policy process, especially at the stages of problem definition, agenda-setting, and policy formulation, requires different parameters than at the implementation stage. Whereas during implementation processes, bureaucracies may gain influence by withholding expert knowledge from their principals, this mechanism is less important at the stages of problem definition and policy formulation. It is not policy-relevant information that is scarce at this stage of the policy process but policymakers' capacity to pay attention to the great amount of information that is fed into the policy process by a multitude of actors. Consequently, scholars studying bureaucratic influence in domestic agenda-setting and policy formulation could gain new insights by conceiving of bureaucracies as attention-seeking organizations, that is, as partially autonomous actors competing with other public and private organizations to supply policy-relevant information to decision-makers. By focusing on a type of bureaucracy whose main tasks are related to the stages of agenda-setting and policy formulation, we described and empirically illustrated two potential pathways through which public administrations may attempt to feed their policy-related knowledge and preferences into the policy process, despite their limited mandates and the comparatively strong control exerted by multiple principals of IPAs.

## References

Abbott, K. W. and Snidal, D. (2010). International Regulation without International Government: Improving IO Performance through Orchestration, *Review of International Organizations* 5 (3): 315–344. DOI: 10.1007/s11558-010-9092-3.

Abbott, K. W., Genschel, P., Snidal, D., and Zangl, B. (eds.) (2015). *International Organizations as Orchestrators*, Cambridge: Cambridge University Press.

Abdenur, A. E. (2021). Climate and Security: UN Agenda-Setting and the "Global South," *Third World Quarterly* 42 (9): 2074–2085.

Alford, J., Hartley, J., Yates, S., and Hughes, O. (2017). Into the Purple Zone: Deconstructing the Politics/Administration Distinction, *American Review of Public Administration* 47 (7). DOI: 10.1177/0275074016638481.

Allan, J. I., Roger, C. B., Hale, T. N., et al. (2021). Making the Paris Agreement: Historical Processes and the Drivers of Institutional Design, *Political Studies* 71 (3): 914–934. DOI: 10.1177/00323217211049294.

Arrow, K. J. (1985). The Economics of Agency. In W. Pratt and R. J. Zeckhauser (eds.), *Principals and Agents: The Structure of Business*, Boston, MA: Harvard Business School Press, 37–51.

Barnett, M. N. and Finnemore, M. (2004). *Rules for the World: International Organizations in Global Politics*, Ithaca, NY: Cornell University Press.

Bauer, M. W. and Ege, J. (2016). Bureaucratic Autonomy of International Organizations' Secretariats, *Journal of European Public Policy* 23 (7): 1019–1037. DOI: 10.1080/13501763.2016.1162833.

Bauer, S. (2006). Does Bureaucracy Really Matter? The Authority of Intergovernmental Treaty Secretariats in Global Environmental Politics, *Global Environmental Politics* 6 (1): 23–49. DOI: 10.1162/glep.2006.6.1.23.

Bauer, S. (2009). The Desertification Secretariat: A Castle Made of Sand. In F. Biermann and B. Siebenhüner (eds.), *Managers of Global Change: The Influence of International Environmental Bureaucracies*, Cambridge, MA: MIT Press, 293–317.

Bauer, S., Busch, P.-O., and Siebenhüner, B. (2009). Treaty Secretariats in Global Environmental Governance. In F. Biermann, B. Siebenhüner, and A. Schreyögg (eds.), *International Organizations in Global Environmental Governance*, London: Routledge, 174–192.

Biermann, F. and Siebenhüner, B. (2009). The Influence of International Bureaucracies in World Politics: Findings from the MANUS Research Program. In F. Biermann and B. Siebenhüner (eds.), *Managers of Global Change: The Influence of International Environmental Bureaucracies*, Cambridge, MA: MIT Press, 319–349.

Biermann, F., Siebenhüner, B., Bauer, S., et al. (2009). Studying the Influence of International Bureaucracies: A Conceptual Framework. In F. Biermann and B. Siebenhüner (eds.), *Managers of Global Change: The Influence of International Environmental Bureaucracies*, Cambridge, MA: MIT Press, 37–74.

Bijker, W. E. and Latour, B. (1988). Science in Action: How to Follow Scientists and Engineers through Society, *Technology and Culture* 29 (4). DOI: 10.2307/3105094.

Busch, P.-O. and Liese, A. (2017). The Authority of International Public Administrations. In M. Bauer, C. Knill, and S. Eckhard (eds.), *International Bureaucracy: Challenges and Lessons for Public Administration Research*, Basingstoke: Palgrave Macmillan, 97–122.

Carpenter, D. P. (2001). *The Forging of Bureaucratic Autonomy*, Princeton: Princeton University Press.

Christopoulos, D. and Ingold, K. (2015). Exceptional or Just Well Connected? *European Political Science Review* 7 (03): 475–498. DOI: 10.1017/S1755773914000277.

Climate Diplomacy (2018). *COP24 Side-Event: Climate, Peace and Security: Progress toward a Preventive Diplomacy*. https://climate-diplomacy.org/events/cop24-side-event-climate-peace-and-security-progress-toward-preventive-diplomacy

Cox, R. and Jacobson, H. (1973). *The Anatomy of Influence*, New Haven, CT: Yale University Press.

Creswell, J. (2009). *Research Design*, Thousand Oaks, CA: SAGE Publications.

Depledge, J. (2005). *The Organization of Global Negotiations: Constructing the Climate Change Regime*, Abingdon; New York: Earthscan.

Depledge, J. (2007). A Special Relationship: Chairpersons and the Secretariat in the Climate Change Negotiations, *Global Environmental Politics* 7 (1): 45–68. DOI: 10.1162/glep.2007.7.1.45.

Dijkstra, H. (2017). Collusion in International Organizations: How States Benefit from the Authority of Secretariats, *Global Governance: A Review of Multilateralism and International Organizations* 23 (4): 601–619. DOI: 10.1163/19426720-02304006.

Dimitrov, R. S. (2016). The Paris Agreement on Climate Change: Behind Closed Doors, *Global Environmental Politics* 16 (3): 1–11. DOI: 10.1162/GLEP_a_00361.

Dubash, N. K. and Rajamani, L. (2010). Beyond Copenhagen: Next Steps, *Climate Policy* 10 (6).

Eckersley, R. (2020). Rethinking Leadership: Understanding the Roles of the US and China in the Negotiation of the Paris Agreement, *European Journal of International Relations* 26 (4). DOI: 10.1177/1354066120927071.

Federal Foreign Office (2018). *United Nations: Germany Initiatives Group of Friends on Climate and Security*. www.auswaertiges-amt.de/en/aussenpolitik/themen/klima/climate-and-security-new-group-of-friends/2125682

Fernandez, R. M. and Gould, R. v. (1994). A Dilemma of State Power: Brokerage and Influence in the National Health Policy Domain, *American Journal of Sociology* 99 (6): 1455–1491. DOI: 10.1086/230451.

Figueres, C. (2013). Climate Policy: A New Foundation of Stability and Prosperity, *Climate Policy* 13 (5). DOI: 10.1080/14693062.2013.822736.

Frederickson, H. G., Smith, K. B., Larimer, C. W., and Licari, M. J. (2018). *The Public Administration Theory Primer*, 3rd ed., New York: Westview Press.

G7 Germany (2015). *Think Ahead. Act Together. An Morgen Denken. Gemeinsam Handeln.* Leaders' Declaration G7 Summit, June 7–8. www.g7germany.de/Content/DE/_Anlagen/G7_G20/2015-06-08-g7-abschluss-eng___blob=publicationFile&v=6.pdf

G20 Australia (2014). *G20 Leaders' Communiqué Brisbane Summit, November 15–16.* www.bundesregierung.de/resource/blob/975254/474688/2943250c014af268b59fc552278ad634/2014-g20-abschlusserklaerung-eng-data.pdf?download=1

Gehring, T. (2012). International Environmental Regimes as Decision Machines. In P. Dauvergne (ed.), *Handbook of Global Environmental Politics*, 2nd ed., Cheltenham: Edward Elgar, 51–63.

Green, J. F. (2014). *Rethinking Private Authority*, Princeton: Princeton University Press.

Greene, B. (2016). *Impossible Isn't a Fact; It's an Attitude: Christiana Figueres at TED2016.* https://blog.ted.com/impossible-isnt-a-fact-its-an-attitude-christiana-figueres-at-ted2016/

Hajer, M. (2003). Policy without Polity? Policy Analysis and the Institutional Void, *Policy Sciences* 36 (2): 175–195. DOI: 10.1023/A:1024834510939.

Hale, T. (2016). "All Hands on Deck": The Paris Agreement and Nonstate Climate Action, *Global Environmental Politics* 16 (3): 12–22. DOI: 10.1162/GLEP_a_00362.

Hall, N. (2016). *Displacement, Development, and Climate Change: International Organizations Moving beyond Their Mandates*, Abingdon; New York: Routledge.

Hall, N. and Woods, N. (2018). Theorizing the Role of Executive Heads in International Organizations, *European Journal of International Relations* 24 (4). DOI: 10.1177/1354066117741676.

Hawkins, D. G., Lake, D. A., Nielson, D. L., and Tierney, M. J. (2006). Delegation under Anarchy: States, International Organizations, and Principal-Agent Theory. In D. Hawkins, D. A. Lake, D. L. Nielson, and M. J. Tierney (eds.), *Delegation and Agency in International Organizations,* Cambridge: Cambridge University Press, 3–38.

Herold, J., Liese, A., Busch, P.-O., and Feil, H. (2021). Why National Ministries Consider the Policy Advice of International Bureaucracies: Survey Evidence from 106 Countries, *International Studies Quarterly* 65 (3).

Hickmann, T. and Elsässer, J. P. (2020). New Alliances in Global Environmental Governance: How Intergovernmental Treaty Secretariats Interact with Non-state Actors to Address Transboundary Environmental Problems, *International Environmental Agreements: Politics, Law and Economics* 20 (3): 459–481. DOI: 10.1007/s10784-020-09493-5.

Jacobs, M. (2016). High Pressure for Low Emissions: How Civil Society Created the Paris Climate Agreement, *Juncture* 22 (4): 314–323.

Jänicke, M. (2017). The Multi-level System of Global Climate Governance: The Model and Its Current State, *Environmental Policy and Governance* 27 (2): 108–121. DOI: 10.1002/eet.1747.

Jinnah, S. (2014). *Post-Treaty Politics: Secretariat Influence in Global Environmental Governance*, Cambridge, MA: MIT Press.

Jordan, A. J., Huitema, D., Hildén, M., et al. (2015). Emergence of Polycentric Climate Governance and Its Future Prospects, *Nature Climate Change* 5 (11): 977–982. DOI: 10.1038/nclimate2725.

Jörgens, H., Kolleck, N., and Saerbeck, B. (2016). Exploring the Hidden Influence of International Treaty Secretariats: Using Social Network Analysis to Analyse the Twitter Debate on the "Lima Work Programme on Gender," *Journal of European Public Policy* 23 (7): 979–998. DOI: 10.1080/13501763.2016.1162836.

Keck, M. E. and Sikkink, K. (1999). Transnational Advocacy Networks in International and Regional Politics. *International Social Science Journal* 51 (159): 89–101.

King, E. (2014). *Pope Francis Backs Global Efforts to Tackle Climate Change.* www.climatechangenews.com/2014/11/27/pope-francis-backs-un-efforts-to-tackle-climate-change

Kinley, R. (2017). Climate Change after Paris: From Turning Point to Transformation, *Climate Policy* 17 (1): 9–15. DOI: 10.1080/14693062.2016.1191009.

Kinley, R., Cutajar, M. Z., de Boer, Y., and Figueres, C. (2021). Beyond Good Intentions, to Urgent Action: Former UNFCCC Leaders Take Stock of Thirty Years of International Climate Change Negotiations, *Climate Policy* 21 (5). DOI: 10.1080/14693062.2020.1860567.

Kok, M. T. J. and Ludwig, K. (2022). Understanding International Non-state and Subnational Actors for Biodiversity and Their Possible Contributions to the Post-2020 CBD Global Biodiversity Framework: Insights from Six International Cooperative Initiatives, *International Environmental Agreements: Politics, Law and Economics* 22 (1). DOI: 10.1007/s10784-021-09547-2.

Kolleck, N., Well, M., Sperzel, S., and Jörgens, H. (2017). Climate Change Education through Social Networks, *Global Environmental Politics* 17 (4). DOI: 10.1162/GLEP_a_00428.

Lagarde, C. (2014). *The IMF at 70: Making the Right Choices: Yesterday, Today, and Tomorrow by Christine Lagarde, Managing Director, IMF.* www.imf.org/en/News/Articles/2015/09/28/04/53/sp101014

Littoz-Monnet, A. (2017). Expert Knowledge as a Strategic Resource: International Bureaucrats and the Shaping of Bioethical Standards, *International Studies Quarterly* 61 (3): 584–595. DOI: 10.1093/isq/sqx016.

Marks, G. (1993). Structural Policy and Multilevel Governance in the EC. In A. Cafruny and G. Rosenthal (eds.), *The State of the European Community. Vol. 2: The Maastricht Debates and Beyond*, Boulder, CO: Lynne Rienner, 391–410.

Mathiason, J. (2007). *Invisible Governance: International Secretariats in Global Politics*, Bloomfield: Kumarian Press.

Mayring, P. and Frenzel, T. (2014). Qualitative Inhaltsanalyse. In N. Baur and J. Blasius (eds.), *Handbuch Methoden der empirischen Sozialforschung*, Wiesbaden: Springer VS, 543–556.

McCubbins, M. D., Noll, R. G., and Weingast, B. R. (1987). Administrative Procedures as Instruments of Political Control, *Journal of Law, Economics, and Organization* 3 (2). DOI: 10.1093/oxfordjournals.jleo.a036930.

Mederake, L., Saerbeck, B., Goritz, A., et al. (2021). Cultivated Ties and Strategic Communication: Do International Environmental Secretariats Tailor Information to Increase Their Bureaucratic Reputation?, *International Environmental Agreements: Politics, Law and Economics* 22 (3): 481–506. DOI: 10.1007/s10784-021-09554-3.

Meuser, M. and Nagel, U. (2009). The Expert Interview and Changes in Knowledge Production. In A. Bogner, B. Littig, and W. Menz (eds.), *Interviewing Experts, Research Methods Series*, Basingstoke: Palgrave Macmillan, 17–42.

Milkoreit, M. (2019). The Paris Agreement on Climate Change: Made in USA?, *Perspectives on Politics* 17 (4): 1019–1037. DOI: 10.1017/S1537592719000951.

Miller Smallwood, J., Orsini, A., Kok, M. T. J., Prip, C., and Negacz, K. (2022). Global Biodiversity Governance: What Needs to Be Transformed? In I. Visseren-Hamaker and M. T. J. Kok (eds.), *Transforming Biodiversity Governance*, Cambridge: Cambridge University Press, 43–66.

Mou, F. (2015). *Churches and Governments in the Region Must Take Action on Climate Change*. www.loopnauru.com/content/churches-and-governments-region-must-take-action-climate-change

Nielson, D. L. and Tierney, M. J. (2003). Delegation to International Organizations: Agency Theory and World Bank Environmental Reform, *International Organization* 57 (2): 241–276. DOI: 10.1017/s0020818303572010.

Niskanen, W. A. (2017). *Bureaucracy and Representative Government*, New York: Routledge.

Olsen, J. P. (2006). Maybe It Is Time to Rediscover Bureaucracy, *Journal Of Public Administration Research and Theory* 16 (1): 1–24. DOI: 10.1093/jopart/mui027.

Ostrom, E. (2010). Polycentric Systems for Coping with Collective Action and Global Environmental Change, *Global Environmental Change* 20 (4): 550–557. DOI: 10.1016/j.gloenvcha.2010.07.004.

Pattberg, P., Widerberg, O., and Kok, M. T. J. (2019). Towards a Global Biodiversity Action Agenda, *Global Policy* 10 (3). DOI: 10.1111/1758-5899.12669.

Raustiala, K. and Victor, D. G. (2004). The Regime Complex for Plant Genetic Resources, *International Organization* 58 (2).

Reinalda, B. and Verbeek, B. (1998). Autonomous Policy Making by International Organizations: Purpose, Outline, and Results. In B. Reinalda and B. Verbeek (eds.), *Autonomous Policy Making by International Organizations*, London: Routledge, 1–8.

Saerbeck, B., Well, M., Jörgens, H., Goritz, A., and Kolleck, N. (2020). Brokering Climate Action: The UNFCCC Secretariat between Parties and Nonparty Stakeholders, *Global Environmental Politics* 20 (2): 105–127. DOI: 10.1162/glep_a_00556.

SCBD (2006). *COP 6 Decision VI/20. Cooperation with Other Organizations, Initiatives and Conventions*. www.cbd.int/decision/cop/?id=7194

SCBD (2010). *Decision Adopted by the Conference of the Parties to the Convention on Biological Diversity at Its Tenth Meeting.* www.cbd.int/doc/decisions/cop-10/cop-10-dec-21-en.pdf

SCBD (2016). *Cancun Declaration on Mainstreaming the Conservation and Sustainable Use of Biodiversity for Well-Being.* www.cbd.int/cop/cop-13/hls/cancun%20declaration-en.pdf

SCBD (2021a). *Elizabeth Maruma Mrema Executive Secretary of the Convention on Biological Diversity 2021 United Nations Biodiversity Conference Opening Statement.* www.Cbd.Int/Meetings/COP-15

SCBD (2021b). *Previous Business Decisions.* www.cbd.int/business/bc/bd.shtml

SCBD (2022a). *CBD Meetings for Business and Biodiversity.* www.cbd.int/business/meetings-events/bm.shtml

SCBD (2022b). *Welcome to the Business Engagement Programme.* www.cbd.int/business/

Schreurs, M. A. and Tiberghien, Y. (2007). Multi-level Reinforcement: Explaining European Union Leadership in Climate Change Mitigation, *Global Environmental Politics* 7 (4).

Sommerer, T., Agné, H., Zelli, F., and Joachim Bes, B. (2022). *Global Legitimacy Crises: Decline and Revival in Multilateral Governance*, Oxford: Oxford University Press.

Thew, H., Middlemiss, L., and Paavola, J. (2021). Does Youth Participation Increase the Democratic Legitimacy of UNFCCC-Orchestrated Global Climate Change Governance?, *Environmental Politics* 30 (6). DOI: 10.1080/09644016.2020.1868838.

Ulloa, A. M. (2022). Accountability as Constructive Dialogue: Can NGOs Persuade States to Conserve Biodiversity?, *Global Environmental Politics* 23 (1): 42–67. DOI: 10.1162/glep_a_00673.

UN (1992a). *Convention on Biological Diversity.* www.cbd.int/doc/legal/cbd-en.pdf

UN (1992b). *United Nations Framework Convention on Climate Change.* https://unfccc.int/resource/docs/convkp/conveng.pdf

UNFCCC (2014). *Momentum for Change Annual Report, 2014.* https://unfccc.int/mfc2014/

UNFCCC (2019). *UN Climate Change Conference COP 25 2-13 December 2019 Madrid, Spain UNFCCC Pavilion – Events Schedule.* https://unfccc.int/sites/default/files/resource/COP25_Pavilion_Schedule.pdf

UNFCCC (2020). *United Nations Climate Change Annual Report 2019.* https://unfccc.int/about-us/annual-report/annual-report-2019

UNFCCC (2021). *Decision-/CP.26 Glasgow Climate Pact.* https://unfccc.int/sites/default/files/resource/cop26_auv_2f_cover_decision.pdf

Van Asselt, H. (2011). Managing the Fragmentation of International Environmental Law: Forests at the Intersection of the Climate and Biodiversity Regimes, *New York University Journal of International Law and Politics* 44: 1205–1278.

Varone, F., Ingold, K., and Fischer, M. (2019). Policy Networks and the Roles of Public Administrations. In Y. Emery, A. Ladner, S. Nahrath, N. Soguel, and S. Weerts (eds.), *Swiss Public Administration: Making the State Work Successfully*, Vol. 18, *Governance and Public Management Series*, Cham: Palgrave Macmillan, 339–353.

Weber, M. (2018). The Profession and Vocation of Politics. In P. Lassmann (ed.), *Weber: Political Writings*, Cambridge: Cambridge University Press, 309–369.

Well, M., Saerbeck, B., Jörgens, H., and Kolleck, N. (2020). Between Mandate and Motivation: Bureaucratic Behavior in Global Climate Governance, *Global Governance: A Review of Multilateralism and International Organizations* 26 (1): 99–120.

Wilson, W. (1887). The Study of Administration, *Political Science Quarterly* 2 (2). DOI: 10.2307/2139277.

Wit, D. de, Ostovar, A. L., Bauer, S., and Jinnah, S. (2020). International Bureaucracies. In F. Biermann and R. E. Kim (eds.), *Architectures of Earth System Governance: Institutional Complexity and Structural Transformation*, Cambridge: Cambridge University Press, 57–74.

Workman, S. (2015). *The Dynamics of Bureaucracy in the US Government: How Congress and Federal Agencies Process Information and Solve Problems*, New York: Cambridge University Press.

World Bank (2015). *World Bank/IMF Annual Meetings 2015: Development Committee Communiqué*. www.worldbank.org/en/news/press-release/2015/10/10/world-bank-imf-annual-meetings-2015-development-committee-communique

Zelli, F. (2018). Effects of Legitimacy Crises in Complex Global Governance. In J. Tallberg, K. Bäckstrand, and J. A. Scholte (eds.), *Legitimacy in Global Governance: Sources, Processes, and Consequences*. Oxford: Oxford University Press, 169–186.

Zelli, F. and van Asselt, H. (2013). The Institutional Fragmentation of Global Environmental Governance: Causes, Consequences, and Responses, *Global Environmental Politics* 13(3).

# 5

# Moving beyond Mandates

*The Role of UNDP Administrators in Organizational Expansion*<sup>*</sup>

NINA HALL

> Climate change threatens to increase the frequency and severity of natural disasters – which have already cost the global economy $2 trillion over the last twenty years.... Small Island Developing States are seeing the encroachment of sea water on their lands and ground water, and are threatened by more intense storms, as we have seen this year in Vanuatu, Bahamas, and Dominica. In drought-prone regions like the Sahel and the Horn of Africa, food insecurity and poor harvests become more frequent.
> – Helen Clark, as Administrator of the United Nations Development Programme (UNDP 2015)

## 5.1 Introduction

Climate change is threatening developing states, as Helen Clark's remarks emphasize, and many do not have the capacity, or finances, to adapt. Adaptation costs for developing countries are large, and global estimates vary between USD 19 billion and USD 429 billion annually by 2050 (Watkiss et al., 2014). Although states have established various new bilateral and multilateral climate adaptation funds, they have not established an international adaptation organization for implementing climate adaptation projects. Rather the assumption is that existing development and humanitarian institutions will integrate climate adaptation into their mandates (Hall 2016b). Yet many international development and humanitarian organizations were established in the aftermath of World War II with no mandate for climate change or the environment. This provokes an important question: How are existing international development institutions responding to climate change? Are they integrating climate adaptation into their mandates?

---

[*] This chapter draws on Hall (2016b).

This chapter focuses specifically on the United Nations Development Programme (UNDP), the United Nation's largest development entity, and examines how and why it moved beyond its original mandate to engage with climate adaptation. In doing so, this chapter takes a different approach from other authors in this book, as it focuses on an international development organization, which is not commonly identified as a core part of the climate change, or any environmental, regime. Although scholars in this book, and elsewhere, have examined the autonomy and effectiveness of international environmental bureaucracies, they have not sufficiently examined how *nonenvironmental* international organizations are addressing climate change and its effects (Biermann and Siebenhüner 2009; Jinnah 2014).[1] It is critical to study how a wide range of institutions are engaging with climate change, as it has implications for health, gender equality, and human rights. Furthermore, UNDP is an important case to study given its presence in over 120 states and influence across the UN system. However, states neither established nor intended UNDP to focus on the impact of or climate change.

This chapter examines how an international service-orientated bureaucracy adapted its own mandate through "self-directed action" (Park and Weaver 2012). It is a relevant comparative case for this book, which largely focuses on the influence of international secretariats, as mandate change in UNDP also involves interstate negotiations at the UNDP Executive Board. The difference is that these negotiations do not produce an international agreement – such as the United Nations Framework Convention on Climate Change (UNFCCC) Paris Agreement – but rather set the direction for UNDP. This chapter suggests that successive executive heads have set their own vision for UNDP, lobbied states, and influenced mandate expansion.

Section 5.2 examines two existing theoretical explanations for mandate change: state-driven and agent-driven. It puts forward an additional elaboration of principal–agent theory: that organizations may not always seek to expand and maximize their scope. Rather, executive heads make decisions about whether and when to expand depending on material, ideational, or normative changes to their external environment.

Section 5.3 traces mandate change in UNDP. It draws on over fifty interviews carried out between 2009 and 2013 with states and international organizations in Geneva (where UNDP has an office), in New York (UNDP headquarters), and at the (UNFCCC Conference of the Parties [COP15] held in Copenhagen in 2009). Interview participants were selected on the basis that they had worked on climate change for the UNDP and/or had a senior role overseeing UNDP's work (as a state

---

[1] An exception is the literature on the World Bank's environmental reforms; see for instance Nielson and Tierney (2003).

representative and/or an international bureaucrat).[2] Interviews were semistructured and lasted for approximately forty to sixty minutes. They were used to identify the timeline of, and reasons for, UNDP's engagement with climate change. Participants were also asked to identify other potential interviewees ("snowball" interviewing) and official documents, which are cited wherever possible throughout this chapter rather than interviews.

The chapter also draws on an extensive analysis of UNDP board meetings' decisions and administrators' speeches from 2000 to 2016. The author searched these public documents and distinguished whether these documents (i) identified climate change as a problem, (ii) mentioned briefly UNDP's role in addressing climate change, and/or (iii) elaborated a substantive role for UNDP in addressing climate change. The author, as directed by interviewees, also examined other significant UNDP strategic documents, evaluations, and reports that elaborated UNDP's climate change policies. These documents, triangulated with the interview transcripts, are the basis for the case study of UNDP's engagement with climate change.

Section 5.4 finds that organizational change, when it occurred, was led by UNDP Administrators, and not by states. It suggests that administrators decided whether and how to expand into a new issue area and then lobbied states to endorse this expansion. This is an important contribution to existing theories of international bureaucracies, which often assume that mandate expansion is state-led or that bureaucracies always seek to expand.

## 5.2 Explaining Mandate Change in International Organizations

States set, adjust, and monitor the mandates of international organizations and are also responsible for funding these organizations to realize their missions. Realist scholars have emphasized the power of hegemonic states in determining international organizations' actions (Mearsheimer 1995). The United States, for instance, nominates the head of the World Bank and has veto power on its executive board and so can directly influence its activities. States may even eject leaders who do not follow their will: The United States, for example, ousted the executive head of the Organization for the Prohibition of Chemical Weapons (OPCW), when he insisted that the OPCW should be allowed to inspect the United States as

---

[2] The author interviewed UNDP staff in the Energy and Environment Group (EEG); Human Development Report Office; the Bureau for Crisis Prevention and Recovery; Bureau for Policy Development; African Adaptation Programme; Small Grants Programme; Gender Equality Team; and the African Regional Team. She had discussions with previous UNDP administrators, Gus Speth, Helen Clark, and Achim Steiner. She also interviewed donor and developing state representatives to UNDP; officials working in the Secretary-General's Climate Change Support Team; and the United Nations System Chief Executives Board for Coordination.

thoroughly as it did other signatories of the Convention on Chemical Weapons (Simons 2013). These are state-driven explanations of international organizations' behavior.

Yet even the most powerful states do not determine all that goes on in an international organization. As principal–agent scholars have noted, states do not typically delegate in detail every task they expect an institution to carry out but rather expect an organization to use its expertise to respond to new issues or circumstances appropriately. Scholars – and practitioners – understand that international organizations have room to interpret their mandates, as the exact parameters are often ambiguous.

In fact, a common assumption of principal–agent theory is that international bureaucracies seek to maximize their autonomy from states. These scholars are interested in so-called agency slack – when an agent (international organization) strays from their principal's (member states) preferences (Hawkins et al. 2006). Scholars have sought to explain why organizations expand by looking at the degree of delegated discretion to an organization, the strength of member state preferences, the degree of consensus within the executive board, voting structures, and the costs of monitoring an organization (Hawkins et al. 2006). States for instance, may establish institutional checks and sanction an institution by controlling the budget and/or overriding decisions. States face a trade-off between the costs of monitoring an agent and the benefits they derive from leaving an agent to implement its mandate more autonomously. Although principal–agent theory offers important insights, scholars tend to assume that international bureaucracies have an inherent interest in maximizing their budget, tasks, and autonomy, as they are "competence-maximizers" (Pollack 2003: 39). In this view, international organizations' preferences are somewhat fixed, and any variation in organizational expansion is determined by the nature of delegation and states' interest or ability to monitor their agents.

Here I suggest that executive heads have a critical role in deciding whether they want to pursue mandate expansion. The preferences of international organizations are not fixed but may evolve in reaction to changes in the external environment. Scholars have already demonstrated how executive heads have influenced the direction of the United Nations High Commissioner for Refugees (UNHCR; Betts 2012), the United Nations Environment Programme (UNEP; Ivaanova 2010), and the World Bank through "self-directed action" (Park and Weaver 2012). In fact, Cox (1969: 205) argues that "the quality of executive leadership may prove to be the most critical single determinant of the growth in the scope and authority of international organizations."

Scholars have also explored variation in executive heads' influence (Woods et al. 2015). Biermann and Siebenhüner (2009: 58), for instance, argue that international environmental bureaucracies are more influential when they have strong leadership, which they define as "charismatic, visionary, and popular as well as flexible and reflexive." Meanwhile, Hall and Woods (2018) have explored how executive heads can overcome legal-political, bureaucratic, and financial constraints. Public opinion, the personality and skills of individual leaders, and the nature of the problem they seek to address may also be important (Park and Weaver 2012). Drawing on this international relations scholarship, I suggest that executive heads deliberate over expansion, and will not always seek to expand and maximize their scope. They will consider the financial opportunities and ideational and normative reasons for expanding.

Firstly, international organizations are reliant on funding to implement their mandates. Furthermore, they operate in an increasingly competitive and complex marketplace with scarce resources. Core funding for UN institutions has decreased over time and donors increasingly favor earmarked financing (funding that is targeted for certain regions, topics, or projects). In 2014, for instance, USD 14.2 billion of the UN system's funding (65 percent) was earmarked while only USD 7.5 billion was allocated to the core budget.[3] Earmarked financing gives executive heads less discretion than core funding. Although there are multiple new sources of financing for multilaterals – private actors (such as Bill and Melinda Gates) and multilateral trust funds (such as GAVI) – there are also an increasing number of development actors trying to secure this funding. Executive heads have considerable scope to look beyond their executive board for funding but must consider the quantity of funding and also the quality (core or earmarked) (Graham 2015). Resource dependency theory scholars would argue that executive heads will be driven by the supply of external material resources (Pfeffer and Salancik 1978). An executive head may decide to take on new issue areas, and expand their organizational mandate, to increase their chances for financing.

In addition, an increasing awareness of how issues are interconnected may facilitate mandate expansion. Scholars have suggested that international organizations will tend to expand in both size and scope as staff "try to square their rationalized abstractions of reality with facts on the ground" (Barnett and Finnemore 2004: 44). This is because "conscientious bureaucrats very quickly recognize that to accomplish a great many ambitious social tasks they need to reach outside the narrow compartments in which we place them" (Barnett and Finnemore 2004: 44). Executive heads, and their staff, are likely to take on new issues – such as climate change – when they see a logical connection to the organization's established expertise. For expansion to occur in this case, executive heads would need

---

[3] UNDP has one of the highest proportions of earmarked budgets in the UN system. Almost 60 percent of their budget was tightly earmarked in 2016 (Schmid, Reitzenstein, and Hall 2021), 446.

to perceive an issue linkage, or causal connection, between their mandate and the new issue (Hall 2016b). For example, scientific research suggests that climate change causes natural disasters, which in turn undermines development efforts. Thus, there is a logical rationale for a development agency to engage with the effects of climate change.

Alternatively, expansion may be driven by normative reasons: Executive heads may see a critical role they should play in a new issue area. Even if an issue linkage is not strong or present, they may look to forge one. This may be the case for many UN agencies that have a normative agenda to protect human rights. For instance, international bureaucrats may argue that climate change has human rights implications because their core concern is to protect human rights, even if there has not yet been independent scholarship outlining these links.

These three factors (financial, ideational, and normative) can be complementary and mutually reinforcing. This chapter advances a nuanced, dynamic understanding of institutional influence, by looking at individual preferences. It does not assume that preferences of states, international institutions, or executive heads are constant across time. It enriches principal–agent literature and sociological institutionalism by examining how leaders navigate external opportunities and constraints (Hall and Woods 2018). It suggests that we must also look at the evolving preferences of executive heads within a changing environment, and leaders may respond differently to these circumstances. It builds on resource dependency theory by suggesting that executive heads are influenced by their environment and can influence it (Pfeffer and Salancik 1978).

This chapter will explore these theoretical arguments in the context of UNDP's expansion into climate adaptation. For the realist argument to hold we would expect to see powerful states instructing – and even delegating – UNDP to work on climate adaptation. In contrast, if executive heads led UNDP's climate adaptation engagement, they would take the lead and lobby member states at the executive board to prioritize climate adaptation. The UNDP Executive Board is chaired by a rotating state representative (the president), and the UNDP Administrator participates but does not have voting powers (United Nations 2011). Identifying exactly who made the first move is challenging – particularly as the author does not have access to the records of bilateral meetings between UNDP and states. Thus, it is hard to ascertain whether member states in private pushed UNDP to engage with climate change. The chapter relies on interviews with state and international bureaucrats, alongside the official and public records, to understand who drove UNDP's climate adaptation work.[4]

---

[4] It is also possible that organizational expansion was driven neither by member states nor the administrator but by UNDP staff.

## 5.3 UNDP and Climate Adaptation

### *UNDP's Evolving Environmental Mandate (1965–1999)*

The United Nations General Assembly established UNDP in 1965 "to enhance coordination of the various types of technical assistance programs within the UN system" and foster economic development in developing countries (Stokke 2009: 187). UNDP was the merger of two UN entities: the Expanded Programme for Technical Assistance and the United Nations Special Fund. At its inception UNDP had no mandate to address any environmental issues in developing country contexts. This is not surprising – after all, the UN environmental agenda began in 1972 with the Conference on the Human Environment in Stockholm, and the concept of sustainable development was elaborated in 1987, in a report of the World Commission on Environment and Development, "Our Common Future" (World Commission on Environment and Development 1987). During its first twenty years UNDP was a development fund that channeled assistance from donors to UN specialized agencies. It was also a program that directly delivered support to 130 governments to build capacity and develop agricultural and industrial sectors and funded natural resource extraction. UNDP supported, for instance, the development of Ghana's gold extraction industry.[5] By the early 1990s it had evolved into a development agency that did "everything" but had no core specialization or focus (Murphy 2006: 232). It had neither a target population (e.g., for the United Nations Children's Fund [UNICEF] these are children) nor a sectoral focus like other UN agencies (World Health Organization and health). In fact, James Gustave Speth (UNDP Administrator, 1993–1999) stated that it "lacked a clear substantive profile, a focus in development policy terms and a profiled strategy" (Klingebiel 1999: 104).

In the 1990s two consecutive UNDP Administrators sought to build UNDP's role in the environment and sustainable development. William Draper (UNDP Administrator, 1986–1993), a former Wall Street banker, was aware that environmental degradation was occurring and would become more of an issue in the future.[6] He also saw the environment as an area for future business for UNDP, as over 100 world leaders met in 1992 at the United Nations Conference on Environment and Development, commonly referred to as the Rio Earth Summit, to discuss environmental issues.[7] At the Rio Earth Summit, states officially launched the Global Environmental Facility (GEF), a new fund dedicated to address global

---

[5] Telephone interview with former UNDP Environment and Energy Group official, June 4, 2012. Note that UNDP also carried out programs in renewable energies in the 1980s. For instance, it supported China in the development of clean coal technology and helped initiate national energy conservation efforts in Peru.
[6] Telephone interview with former UNDP Environment and Energy Group official, June 4, 2012.
[7] Telephone interview with former UNDP Environment and Energy Group official, June 4, 2012.

environmental issues (Mingst and Karns 2016: 216). The GEF channeled grants from developed to developing states to address biodiversity, climate change, ozone layer depletion, and international waters (Young 2002). UNDP, UNEP, and the World Bank worked together to establish the legal and constitutional framework for the GEF.[8] At the outset, UNDP was one of only three agencies that had access to the GEF, worth USD 1.2 billion (Murphy 2006).

The creation of the GEF signaled the beginning of UNDP's work on climate change, in contrast to its prior main focus on energy demand, supply, and conservation. UNDP, through the GEF, had access to a stream of financing to develop environmental activities, separate from member state contributions. It subsequently used the GEF to assist developing countries to fulfill their requirements to the UNFCCC on mitigation – so-called enabling activities. During his time in office, Draper established UNDP's first environmental unit: the Energy and Natural Resources Unit, which became the main UNDP interlocutor with the GEF.

In 1993, Draper was replaced by Speth. Speth, an active environmentalist, had played a central role in world environmental conferences, including the 1992 Rio Conference, and was a founder of the World Resources Institute. He sought to integrate the concept of sustainable development into UNDP, drawing on the 1987 "Our Common Future" report. In Speth's first speech to all UNDP staff, he outlined his concept of sustainable human development.[9] He arrived with a clear vision that environmental issues were an important priority for UNDP to grapple with and encouraged member states to reorientate its mandate. In 1996 UNDP's executive board endorsed this new vision and the UNDP mission placed "sustainability" at its center (Klingebiel 1999: 106–107). This was a significant reorientation of UNDP and built on the success of the United Nations Conference on Environment and Development in making environmental issues mainstream. Sustainability is a broad concept, and notably Speth did not focus UNDP's mandate specifically on the impacts of climate change in developing countries. This is not surprising as in the 1990s many were still debating whether and how to reduce greenhouse gas emissions, and few policymakers focused on adaptation (Hall 2016b).

Speth renamed the Energy and Natural Resources Unit the Sustainable Energy and Environment Division and focused its energies on securing GEF projects and developing policies on sustainable development. The GEF work dominated the division and brought in the vast majority of its funds. Between 1994 and 1997 UNDP received more than USD 150 million from the GEF, three times the core funding of its programs (Murphy 2006: 271). However, the environment division

---

[8] Telephone interview with former UNDP Environment and Energy Group official, June 4, 2012; Interview with UNDP Environment and Energy Group official, May 22, 2012, New York.
[9] Telephone Interview with former UNDP administrator James Gustave Speth June 14, 2012.

had little engagement or influence over the rest of UNDP's programs as it did not operate with core funding (Young 2002: 85). The Sustainable Energy and Environment Division could operate regardless of support from the executive board as it had a separate stream of accessible funding earmarked for the environment. Thus, by the late 1990s a number of environmental projects were developed and implemented, some of which focused on climate change mitigation, but these were not part of UNDP's mandate or strategic objectives. This shift occurred as a result of the new financing opportunities and Draper's and Speth's acknowledgment of the importance of sustainable development for UNDP.

### Mark Malloch-Brown and the Environment (1999–2005)

In 1999 Lord Mark Malloch-Brown, a former World Bank vice-president, became UNDP Administrator. UNDP was in a financial crisis: The organization's core funding was low. Malloch-Brown (2011: 119) put it simply: "UNDP was poor and the World Bank was rich." He diagnosed that UNDP had attempted to compete across too many areas of development, stretching its expertise too thin. It had agriculture and education experts, public health and forestry units, urban planning expertise and much more, even though the United Nations had other specialized agencies in each of these areas (Malloch-Brown 2011: 121). In response, he sought to create "a highly focused" (UNDP 2001: 2) organization and downgraded the environment and natural resource management "as having little to contribute to the core UNDP mandates of poverty and governance" (UNDP Evaluation Office 2010: vii).

Malloch-Brown made major structural changes to the Environment and Energy Group. He disbanded the forestry program, reduced the number of staff working on the environment, and decentralized the Environment and Energy Group. The UNDP Administrator also discontinued positions in sustainable livelihoods, transport, and sustainable development (UNDP Evaluation Office 2008: 11). He sought to reduce the number of staff in the Bureau for Policy Development, of which the Environment and Energy Group was part, from 250 to fewer than 120 staff members at headquarters, with 98 staff redeployed to the field by 2001 (Malloch-Brown 2011: 2). Malloch-Brown's decentralization and restructuring caused a sharp decline in the number of environment staff positions at headquarters, and a number of the senior environment and energy staff left after his arrival.[10]

UNDP lost much of its environmental policy work, and its environmental staff would be asked at non-GEF environmental meetings, "why are you here?"[11] Yet

---

[10] Telephone interview with former UNDP Environment and Energy Group staff member, October 18, 2010.
[11] Telephone interview with former UNDP Environment and Energy Group official, June 4, 2012.

while the rest of UNDP was facing major cuts, the GEF unit continued to have access to earmarked financial resources for the environment through the GEF. In 2001 UNDP's environmental activities were almost exclusively GEF-funded and there was no core funding (from the executive board) for climate change activities.[12] Staff were encouraged to develop the maximum possible number of projects likely to be approved by the GEF.[13] Thus, UNDP's environment and energy portfolio became even more dependent on the GEF. Malloch-Brown supported climate change activities as long as they were financially self-sustaining, through the GEF or other multilateral funds, and did not drain core resources. Although UNDP did develop some climate mitigation projects during this time, they were not aligned with Malloch-Brown's strategic focus on poverty reduction and good governance.[14]

The UNDP board was broadly supportive of Malloch-Brown's downsizing of the Environment and Energy Group. In 2000 the executive board decided to discontinue environment as a core priority within UNDP's multiyear funding frameworks (UNDP Evaluation Office 2012). Although the environment was reinstated as a priority in 2002, it did not have "status as a core priority supported by core funds" (UNDP Evaluation Office 2012: 12). There was almost no mention of climate change in the official summary of decisions adopted by the executive board between 2000 and 2005, which suggests that climate change was a priority neither for states nor for UNDP officials. Climate change in UNDP did not need to be a high organizational priority as it received funding through the GEF, during a period of declining core contributions to UNDP. UNDP's overarching goal was poverty reduction, and other agencies, such as UNEP, had a greater mandate and expertise in the environmental sphere (Executive Board of the UNDP/UNFPA 2004; UNDP Evaluation Office 2012: 12). In short, under Malloch-Brown UNDP deprioritized the environment and climate change, and this area of work continued only because of the GEF funding.

## *Kemal Derviş and Climate Change (2005–2009)*

In 2005 Kemal Derviş took over as UNDP Administrator. He was a former Turkish minister of economic affairs and had previously worked as the World Bank's chief economist. According to one staff member, he was a "very intellectual and solid

---

[12] Telephone interview with former UNDP Environment and Energy Group official, June 4, 2012.
[13] Telephone interview with former UNDP Environment and Energy Group official, June 4, 2012.
[14] Reliance on GEF funding for environment and energy initiatives meant that global environmental issues took precedence over national objectives and concerns such as pollution and water supply. The 2008 UNDP Evaluation Office was particularly critical of the GEF unit in UNDP for this reason. See UNDP Evaluation Office (2008).

economist and always demanded that policy decisions be rigorously backed by empirical evidence whenever possible."[15] This may have influenced his outlook on climate change, particularly at a time when many governments were reluctant to admit climate change was anthropogenic and did not see climate adaptation as a development priority. When Derviş arrived, UNDP was still in the process of elaborating its positions on climate change.[16] Derviş spent a "considerable amount time guiding the organization" and his senior managers through intellectual discussions to develop a UNDP position.[17] He demanded of his staff rigorous analysis on how mitigation and adaptation issues would impact development trajectories and how the burden of climate change, including financing, should be shared among states.[18] During his first year in office, while these internal discussions took place, he delivered no major speeches on climate change.

In fact, an evaluation of the Environment and Energy Group stated that the environment was not a "core priority for the new administrator" and was critical of the climate related activities taking place (UNDP Evaluation Office 2010: ix). The evaluation argued that "the fit between UNDP's poverty reduction and the GEF objective of mitigating global climate change has been less than convincing" (UNDP Evaluation Office 2010: x). It argued that adaptation was a "more natural area for UNDP to engage in than mitigation, where the benefits are largely global" (UNDP Evaluation Office 2010: x). It noted that there was a high level of dependence on the GEF and emphasis on "going after available money rather than allocating core resources to sets of activities that are consistent with the UNDP mandate" (UNDP Evaluation Office 2010: x). The Environment and Energy Group continued to be supported and driven by the GEF's priorities. Derviş was not initially a strong advocate for UNDP engaging with climate change and neither were states. In the summary of decisions taken by the UNDP Executive Board between 2005 and 2007, there is no mention of climate change.

In 2006 Derviş, and his associate administrator Ad Melkert, began to speak about climate change as a development issue. They both highlighted a new UNDP initiative – the Millennium Development Goals Carbon Facility, which offered developing states financing for carbon emission reductions. Derviş (2006) explained that it was "formulated to assist developing countries in addressing the challenge of climate change while at the same time using carbon financing opportunities to generate alternative and additional financing for reaching the Millennium Development Goals." These speeches correlated with an increased global, and mainstream, interest in climate change. In 2006, the UK government released the

---

[15] Interview with UNDP official, October 7, 2010, New York.
[16] Interview with UNDP official, October 7, 2010, New York.
[17] Interview with UNDP official, October 7, 2010, New York.
[18] Interview with UNDP official, October 7, 2010, New York.

*Stern Review on the Economics of Climate Change*. The report was extremely influential and highlighted the detrimental effect of global warming on the world economy. It claimed that climate change was the greatest and most wide-reaching market failure ever seen and that climate change represented a "grave threat to the developing world and a major obstacle to continued poverty reduction across its many dimensions" (Stern 2006: vii). In addition it noted, "Adaptation will cost tens of billions of dollars a year in developing countries alone, and will put still further pressure on already scarce resources. Adaptation efforts, particularly in developing countries, should be accelerated" (Stern 2006: vii). In 2007 the release of the fourth IPCC report, the UNFCCC summit in Bali (COP13), and Al Gore's movie, *An Inconvenient Truth*, motivated further global awareness of climate change.

Derviş responded to the increased global awareness of climate change and its impact on developing countries. He stated that "global warming can't be looked at as an environmental issue anymore: it is undoubtedly a threat to human development as a whole. All development strategies must therefore account for climate-related risk" (Derviş 2007b). He made climate change a central part of his speech to the executive board in 2007, where he outlined UNDP's three-pronged approach to climate change. This involved: (i) mainstreaming climate change into UNDP's core activities; (ii) creating conditions that allow markets and the private sector to "provide effective solutions to sustainable development and climate change mitigation" (Derviş 2007c); and (iii) increasing the capacity of developing countries to incorporate resilience into national plans. The speech was important as it signaled to the executive board the central importance of climate change to UNDP.

In 2007 UNDP also published its first major report linking climate change to human development. The Human Development Report *Fighting Climate Change: Human Solidarity in a Divided World* (2007: vi) argued that climate change was a development issue as "development progress is increasingly going to be hindered by climate change. So we must see the fight against poverty and fight against the effects of climate change as interrelated efforts." The report forged a strong conceptual linkage between climate change and human development, stating that climate change threatens human development by eroding "human freedoms and limiting choice" (UNDP 2007: 7). It emphasized that developing countries would be the worst hit by climate change, and had the lowest carbon footprints, and thus the international community should assist them in adaptation. It also stated that "human development itself is the most secure foundation for adaptation to climate change" (UNDP 2007). The Human Development Report's main contribution was to identify how climate change would impact on the poorest and those in the Global South.[19] This report was a "very important catalytic moment" according to

---

[19] Interview with former UNDP Human Development Report official, June 12, 2012, Oxford.

one UNDP staff member, as until then climate change was not well understood in UNDP outside of the Environment and Energy Group.[20]

Derviş continued to emphasize the links between development and climate change in 2007, 2008, and 2009. He stated that "should the pace of [climate] change accelerate further, development and adaptation could well become synonymous" (Derviş 2007a). Immediately before the Bali UNFCCC summit Derviş (2007d) published an Op-Ed on climate change and development where he stated that a "failure to act on climate change will have grave consequences for human development in some of the poorest places in the world and it will undermine efforts to tackle poverty." He used public speeches to position UNDP as an agency with expertise in climate change and development. He showcased the UNDP's expertise in climate assistance to member states at the executive board, stating that it was "one of the largest sources of technical assistance for climate change related actions in the world, with an on-going portfolio of about US $2 billion [from GEF]" (Derviş 2007c).

In addition to the Human Development Report, UNDP published a Climate Change Strategy in 2008 aimed at staff and member states. It outlined how to integrate climate change across UNDP and justified why UNDP was the best-positioned agency to work on climate change within the UN system. The strategy built on the Human Development Report, stating that UNDP's overarching goals were "to align human development and climate change management efforts by promoting mitigation and adaptation activities that do not slow down but rather accelerate socio-economic progress" (UNDP 2008a: 7). This goal would be realized through mainstreaming climate change in UNDP's development policies as well as through the UN, national, regional, and international programs and policies. Climate change mainstreaming within UNDP would be led by the Environment and Energy Group and a "cross-practice steering group" of governance, poverty reduction, capacity development, and gender experts who would develop programming tools in each area (UNDP 2008a: 20). The Climate Change Strategy and Human Development Report elaborated an issue linkage between climate change and human development and thus a rationale for UNDP's engagement with climate change.

In 2008 member states officially endorsed UNDP's new role in climate change adaptation and mitigation. UNDP renegotiated with the board its multiyear strategy, to replace the previous multiyear funding framework for 2004–2008. The resulting strategy document (UNDP 2008b) listed four key sectors where UNDP had a mandate to deliver policy advice and technical assistance. These were poverty reduction, democratic governance, crisis prevention and recovery, and

---

[20] Interview with UNDP Energy and Environment Group senior official, May 22, 2012, New York.

environment and sustainable development. "Promoting climate change adaptation" was listed as a subset of the UNDP's environment and energy areas (UNDP 2008b: 34–35). Member states endorsed the strategic plan and thus gave UNDP a clear and official focus on adaptation. Up until then UNDP had no adaptation service line (UNDP 2008b: 34–35).

Although UNDP was mandated to work on climate adaptation it was not always highly prioritized by the organization. Meanwhile, during this period "funding exploded" for climate change adaptation outside of the GEF.[21] New international funds for climate change adaptation emerged, and demand from recipient countries multiplied.[22] This funding provided a strong incentive for UNDP to expand its climate change portfolio between 2008 and 2011. From 2008 onward there was a marked increase in the number of staff working on climate change in the Environment and Energy Group, outside of the GEF-financed projects. Bilateral donors funded UNDP to establish new programs on adaptation, deforestation, and carbon financing.[23] UNDP crated new teams at headquarters to manage these programs. In 2008, for example, the Japanese government gave UNDP USD 92.1 million to implement adaptation programs in twenty African states between 2008 and 2012. This was a major grant that UNDP, in partnership with the United Nations Industrial Development Organization (UNIDO), UNICEF, and the World Food Programme, used to establish the African Adaptation Programme, its largest adaptation program at the time. UNDP established a team to manage the program in Senegal with oversight provided by its headquarters in New York.[24] One staff member explained that "there's a lot more climate change capacity in the Environment and Energy Group and less and less in the other areas."[25] The Environment and Energy Group shifted from being predominantly GEF-reliant to a more even split between GEF and other funding.

In addition, divisions outside of the Environment and Energy Group began to establish their own climate change experts. The gender unit, for instance, established a team of three people to develop policies to link gender and climate change and advocate for gender equality in the UNFCCC negotiations and the

---

[21] Interview with UNDP Environment and Energy Group senior official, May 22, 2012, New York.
[22] Interview with UNDP Environment and Energy Group senior official, May 22, 2012, New York.
[23] UNDP, FAO, and UNEP initiated the UN-REDD program in 2008. REDD, or REDD+, is a financial incentive mechanism under the UNFCCC and stands for Reduction of Emissions due to Deforestation and forest Degradation in developing countries. The UN-REDD program is funded by four major donors including the Norwegian government. The UNDP staff comprises twelve technical and policy staff at the global level (based in New York, Oslo and Geneva), two regional technical advisers for the Asia-Pacific region, and one regional adviser each in Latin America and the Caribbean and Africa. The program is worth around USD 119 million. Interview with UNDP Environment and Energy Group official, October 21, 2010, New York.
[24] UNDP, Africa Adaptation Programme website, www.undp-aap.org/. Interview with UNDP Environment and Energy Group official d, October 6, 2010, and May 21, 2012, New York. Telephone interview with UNDP official, June 1, 2012.
[25] Interview with UNDP official, October 7, 2010, New York.

climate funds.[26] UNDP also established a climate change focal point system at the regional headquarters and country office level. Each regional center was assigned several "qualified people" on climate change.[27] Staff expertise shifted from "environment and energy, to climate change, and now to climate change mitigation and adaptation separately."[28] The Environment and Energy Group sought to train "almost every single staff member from UN resident coordinators to the environmental coordinators on the UNFCCC negotiations and carbon financing."[29] UNDP's expansion into climate change included a reorientation of staff expertise and was enabled by increased climate finance. Derviş' views on whether and how UNDP should engage with climate change evolved over his tenure, in reaction to the changing financial environment, increased demand for normative leadership, and growing awareness of how climate change affected developing countries.

### *Helen Clark and Climate Change (2009–2017)*

In 2009 Helen Clark, a former prime minister of New Zealand, became the new administrator of UNDP. Clark arrived the year of the high-profile UNFCCC summit in Copenhagen. She stated from the outset that climate change should be one of UNDP's top priorities, alongside the Millennium Development Goals. In her first speech to the executive committee in April 2009 she argued that it is "critical" to bring in the "climate change challenge into the center of the way in which we think about development" (Clark 2009b). Clark's position built on Derviş': She reiterated that climate change undermined development efforts and hit the poorest worst. In addition, she outlined a role for UNDP as the "UN agency with a climate and development mandate" as it had "significant expertise in the areas of climate change and sustainable development" (Clark 2009d). Clark (2009d) had a clear view of UNDP's priorities within its mandate: UNDP was mandated to work in four areas; two of these – promotion of democratic governance and crisis prevention and recovery – were stepping stones to their other priorities, poverty reduction, the Millennium Development Goals, and environment and sustainable development.

Clark positioned UNDP as the UN climate change and development agency, partly with the hope of securing additional funding. She viewed the Copenhagen

---

[26] Interview with UNDP gender official h, October 21, 2010, New York. The gendered impacts of climate change were included in a number of speeches by the administrator and the deputy administrator. See Melkert (2008).
[27] Interview with UNDP official, October 12, 2010, New York.
[28] Interview with UNDP official, October 12, 2010, New York.
[29] Interview with UNDP Environment and Energy Group senior official b, May 22, 2012, New York.

summit as an opportunity to establish new climate funds,[30] and she lobbied governments to reduce emissions and commit new, additional, resources to cover adaptation costs of developing countries (Clark 2009e). She argued for a "development deal" at Copenhagen, which would benefit developing countries as well as UNDP. She stated:

> What could be achieved at Copenhagen, including through finance mechanisms being worked on, has significant implications for development. These mechanisms could become a major new and additional source of development financing, complementing, and at some point possibly even surpassing the significance of ODA [overseas development assistance]. A new development paradigm could be in the making.
> 
> *(Clark 2009a)*

She maintained that UNDP should have a role to play in dispersing this new climate financing, agreed upon by states at Copenhagen and Cancun UNFCCC summits.

Clark secured member state support to make climate change a top organizational priority. At the 2009 executive board meeting she stated, "Making the links between Millennium Development Goals achievement and sustainable development has also led me to prioritize UNDP's support to program countries on climate issues and the ongoing negotiations for a new agreement. Development and the impact of climate change and variability cannot be treated as distinct issues. They are inextricably linked" (UNDP 2009). UNDP (2009) also reported on its climate change adaptation expenses (a total of USD 11.7 million) for the first time. In the 2009 Annual Report, Clark asserted that combating climate change was one of UNDP's top mandated priorities (UNDP 2009: 15). This was a remarkable claim to make and a significant shift from UNDP's position in 2000. Donors in 2010 endorsed its position and "called upon UNDP to continue playing a central role in linking climate change to development and helping developing countries to take mitigation and adaptation measures" (UNDP Executive Board 2010: 3). Member states were overall supportive of UNDP's engagement with climate adaptation in the publicly available executive board documentation, but there is no evidence that they initiated this shift.

Significant structural change also occurred under Clark between 2009 and 2011. In 2009 UNDP outlined the need for a "surge" in staff capacity to its executive board (Melkert 2009). Associate Administrator Melkert (2009) argued that climate change was an area of "extraordinary demand" due to preparations for Copenhagen and the hoped-for future agreement on mitigation and adaptation. He stated that there will be "with no doubt the need for substantial extra capacity

---

[30] She highlighted this in conversation with the author at the UNFCCC summit, Copenhagen, 2009. She later stated, "Where I want more focus and action now is on … environment and sustainable development. This is particularly important … as the climate change negotiations enter an intensive period with considerable potential to benefit development" (Clark 2009c).

to support in particular the least developed countries and small island states" (Melkert 2009). UNDP established climate focal points in the regional bureau at headquarters and at regional and country levels. At headquarters each regional bureau established a climate change focal point. For example, the Regional Bureau for Africa had a climate change advisor reporting directly to the bureau's director.[31] UNDP also sent twenty-six climate change focal points to country offices in least developed countries.[32] This was part of a concerted effort to put more staff on the ground, develop climate change programs, and mainstream climate change across UNDP's work. The creation of these new positions represented a significant investment of resources and locked in previous rhetoric and policy changes. These staff changes institutionalized climate change as a central priority for UNDP.

Clark also lobbied states to increase their climate financing at annual board meetings and international summits – from the UNFCCC in Warsaw, Poland (December 2013), to the United Nations Conference on Small Island Developing States in Apia, Samoa (September 2014). She advocated for financing for "climate-integrated development strategies" and state commitment to operationalize and adequately finance the Green Climate Fund (Clark 2012b). She argued that more was needed to meet states' commitments at Copenhagen to raise USD 100 billion annually by 2020, as only USD 50 million had been pledged for seed funding to the Green Climate Fund. She reiterated that climate financing should be additional to current development financing.

In parallel to these changes, UNDP also expanded its adaptation operations considerably. In the early 2000s it had no adaptation projects but by October 2013 it had 193 underway (UNDP 2013; UNDP EEG 2013a, b).[33] In fact, in 2012 Administrator Clark highlighted that UNDP was "the largest implementer of programmes in the UN development system, with more than US $500 million in annual delivery," which translated to support for 140 countries to address climate change in 2011 (Clark 2012a). UNDP's adaptation projects were mainly funded through two sources: the multilateral climate funds (namely, the Special Climate Change Fund, the Least Developed Countries Fund, GEF's strategic priority on adaptation, and the Adaptation Fund) and the Japanese African Adaptation Programme. Under Clark's leadership UNDP established a strong issue linkage between climate change adaptation and human development and secured state support to refocus the organization's efforts on climate change. This was facilitated by an increase in climate finance for mitigation and adaptation.

---

[31] Interview with UNDP Climate Change Focal Point in Regional Bureau for Africa, October 21, 2010.
[32] Interview with UNDP Climate Change Focal Point in Regional Bureau for Africa, October 21, 2010.
[33] UNDP also developed an on-line database of all their adaptation projects: the 'Adaptation Learning Mechanism', www.undp-alm.org/

## 5.4 Explaining Organizational Expansion

The United Nations Development Programme was not established with a mandate to work on climate adaptation, sustainable development, or the environment. Moreover, states did not explicitly instruct UNDP to address climate adaptation as state-driven explanations of organizational change might expect. There is no evidence of states' desire to shift UNDP into climate adaptation in the publicly available official records of the executive board meetings. Furthermore, no interview candidates highlighted the role of states in encouraging UNDP to engage with climate adaptation. Rather, the evidence suggests that successive UNDP Administrators reinterpreted and expanded UNDP's mandate for normative, financial, and ideational reasons. It is worthwhile briefly examining how this external environment evolved in the 1999 to 2015 period, before comparing the particular responses of individual executive heads.

*Financing* for climate change began with the GEF, and was initially targeted solely at mitigation. It was only in the 2000s that grant financing specifically for adaptation was available. In 2000 at the sixth annual UNFCCC summit, as the negotiations over Kyoto became difficult, the European Union agreed to establish an annual climate change fund of USD 15 million to target adaptation as well as mitigation. Subsequently at the next COP in Marrakech in 2001, three multilateral funds were established: the Special Climate Change Fund, based on voluntary donations to facilitate technology transfer from developed to developing states; the Least Developed Countries Fund for least developed countries to develop National Adaptation Programmes of Action; and the Adaptation Fund, which was financed by a 2 percent levy on the Clean Development Mechanism. The establishment of these three climate funds offered new financing opportunities in adaptation as well as mitigation. Then at the UNFCCC in Copenhagen in 2009 states agreed to establish a new Green Climate Fund. They also pledged to mobilize USD 30 billion in total (USD 10 billion per annum) by 2012 and up to USD 100 billion by 2020 for both mitigation and adaptation. As of June 2017 states had pledged USD 10.4 million to the Green Climate Fund from forty-three countries.[34] Thus by the mid-2000s there were strong financial incentives for development institutions to work on climate adaptation, and these incentives became stronger over time as more financing was pledged and delivered.

In terms of *ideational* links, in the 1990s climate change was seen almost exclusively as an environmental issue (much like the ozone hole and its Montreal Protocol). It was only in the mid-2000s that academics, developing countries, and nongovernmental organizations (NGOs) highlighted how climate change would have disastrous impacts on developing countries, undermining their chances of

---

[34] www.greenclimate.fund/how-we-work/resource-mobilization

development. As noted earlier, the 2006 Stern Review was a key part of building this connection between climate change and the economy. Development NGOs, developing countries, and other experts built strong issue linkages between development and adaptation over the early 2000s. By 2009 it was more commonly accepted that climate change would impact not just polar bears but also people in developing countries (Hall 2016b). Furthermore, adaptation became increasingly intertwined with development efforts. Thus by 2009 there were strong ideational reasons for development actors to develop policies to address adaptation.

Alongside this, there was also greater global, mainstream, awareness of the moral urgency of climate change. It is difficult to pinpoint an exact moment: Perhaps it is 2005 when climate change was one of the top agenda items at the G8 summit in Gleneagles.[35] From then on climate change became a regular agenda item on the G7/8 and G20 agenda. It could also be dated to the release of Al Gore's movie, An Inconvenient Truth, in 2006, which brought the perils of climate change to a broad global audience. There was also significant public and political debate about the cause and the scale of climate change as well as the appropriate global and national policy responses. Nevertheless, by the mid-2000s climate change was accepted by many states as a major global challenge, and it is no surprise that UN institutions felt a need to respond. In fact, the UN Secretary-General, Ban Ki-Moon, requested all UN agencies to establish change focal points and develop a united UN climate policy in the lead-up to Copenhagen. The United Nations Secretariat launched an initiative to mitigate their emissions, "Greening the Blue."[36]

Reflecting on these three external factors – financing, ideational, and normative – we would expect UNDP Administrators to expand into climate adaptation from the mid-2000s onward. Previous to that there was little incentive for UNDP to take on climate adaptation – as there was no ideational, monetary, or normative rationale. However, there were financial, ideational and normative reasons to engage with a broad range of environmental issues under the umbrella of sustainable development. As we saw, Draper, a venture capitalist, saw that environmental change was an important global challenge (normative), which related to development concerns (issue linkage), particularly in the lead-up and aftermath of the 1992 United Nations Conference on Environment and Development. Draper also saw financial opportunities for UNDP through the GEF. His successor, Speth, was a strong environmentalist and elaborated a vision for UNDP in sustainable development (normative) and in doing so connected environmental concerns to its development mandate (ideational). UNDP continued to work through the GEF on

---

[35] One could also argue we are seeing another wave of mainstream awareness of climate change with today's #FridayforFutures strikes, initiated by Greta Thunberg, and also the Extinction Rebellion protests. See Hall (2016a and 2022)
[36] www.greeningtheblue.org/

climate mitigation and other environmental issues. By 1996 it had a mandate to work on sustainable human development, but not explicitly on climate adaptation.

However, UNDP's new environmental mandate was not set in stone. The following administrator, Malloch-Brown, deprioritized the environment in the late 1990s and early 2000s as he did not see a core role for UNDP in this issue area (issue linkage) and was concerned that the organization had spread itself too thinly across many issue areas. As a result, there was little engagement with climate change between 2000 and 2007, beyond UNDP's implementation of GEF projects. This is an excellent example of how executive heads may interpret their external environment differently and limit their mandates accordingly.

In contrast, Derviş and Clark both prioritized climate change and spoke frequently about the connection between climate change and human development. They were driven by normative, ideational, and financial opportunities. Derviş echoed a growing view at the time that climate change would have disastrous effects on the economies of developing countries who were most vulnerable (normative and ideational). Under his term UNDP published the 2007 Human Development Report, the Climate Change Strategy, and the 2008 UNDP Strategy, which made climate change an institutional priority. UNDP's expansion into climate change operations was also enabled by an expansion of financing opportunities from multilateral trusts and from bilateral donors. It is unlikely UNDP would have invested so many staff resources or developed almost two hundred adaptation projects if new climate financing was not available. It is significant that much of this financing was earmarked, and often from multilateral trust funds, and was not core funding from the executive board. In 2010 member states endorsed climate change as a mandated goal for UNDP. Clark also outlined a strong normative role for UNDP as the "UN agency with a climate and development mandate" based on an issue linkage between climate change and sustainable development. She also saw great financial opportunities for UNDP through new climate financing mechanisms. Overall, UNDP Administrators set the strategic direction of UNDP and official documentation suggests that the executive board tended to follow their lead. The board endorsed Malloch-Brown's shifting away from the environment and Derviş and Clark's prioritization of climate change.

This chapter found that UNDP Administrators developed their own visions for UNDP's role in addressing climate change. These views evolved as normative, ideational, and financial opportunities changed, and based on their own assessment of UNDP's role in global governance, as this chapter has traced. A more fine-grained analysis of the early years of UNDP's work and interviews with more members of the former staff could explore whether particular UNDP staff, or NGOs, influenced UNDP Administrators' positions on climate change and the environment. Normative entrepreneurs within or outside UNDP could have driven mandate expansion. There is certainly evidence in other international institutions that senior managers and innovative bureaucrats played a strong role in driving organizational expansion. Some staff in the International Organization for Migration (IOM), for instance, pushed the issue of climate change and migration, even when IOM's Director-General was not fully engaged (2016b). Another potential explanation is that staff may have steered UNDP away from climate adaptation. This was the case in UNHCR, where many staff were initially reluctant to engage in debates over so-called "climate refugees" (Hall 2016b). Further research should also focus on how states' views of UNDP's role changed over time. It would be useful, for example, to interview all the UNDP board members over the 2000–2015 period and have access to their internal records of bilateral meetings with UNDP Administrators.

## 5.5 Conclusion

Executive heads of international bureaucracies play an important role in determining whether and how to expand an organization's mandate. The chapter found that two administrators played a central role in shifting UNDP toward the *environment* (Draper and Speth), and two others in prioritizing climate adaptation (Derviş and Clark). However, one administrator (Malloch-Brown) deprioritized environmental issues within UNDP. Leaders who favored expansion often did so when they saw a confluence of normative, ideational, and financial reasons. During Derviş's and Clark's leadership, climate change became an increasingly accepted and important global concern (normative). They also saw an issue linkage between climate change and UNDP's development mandate (ideational). It was not enough to see climate change as an urgent global issue if they did not see a role for UNDP in addressing it (this was Malloch-Brown's position). Derviş and Clark often referred to this issue linkage in their speeches, and justified UNDP's expansion. Thirdly, UNDP was able to support developing countries with climate adaptation because of new multilateral and bilateral climate funds (financing). This created a strong incentive to develop expertise on adaptation. Overall, this chapter emphasized the importance of executive heads in mandate expansion. In particular, it suggested that how UNDP Administrators perceive the financial, ideational, and normative opportunities will influence their decision to expand into a new issue area.

Further research is needed to increase the generalizability beyond UNDP and climate adaptation. Scholars could look at why and how other international organizations expanded into climate change. Scholarship exists on the expansion of UNHCR and IOM but not on other important organizations such as UNICEF, UN-WOMEN, the World Health Organization, the World Bank, and the Office of the United Nations High Commissioner for Human Rights. Comparison should also be extended to how international organizations have engaged with other issue areas such as gender equality or indigenous rights. We also need a broader understanding of global adaptation governance and in particular what role the UNDP Administrator and the UN Secretary-General played in encouraging other UN entities to engage with adaptation (Hall and Persson 2018: 540–566). After all, the UNDP Administrator chairs the United Nations Development Group, which gathers thirty-two UN funds, programs, specialized agencies, and other bodies that work to support sustainable development.[37] Staff of international organizations may not only expand their own organizational mandates but also influence bureaucrats in other institutions to expand theirs.

## References

Barnett, M. N. and Finnemore, M. (2004). *Rules for the World: International Organizations in Global Politics*, Ithaca, NY: Cornell University Press.

Betts, A. (2012). UNHCR, Autonomy and Mandate Change. In J. E. Oestrich (ed.), *International Organizations as Self-Directed Actors*, Abingdon: Routledge, 118–140.

Biermann, F. and Siebenhüner, B. (eds.) (2009). *Managers of Global Change: The Influence of International Environmental Bureaucracies*, Cambridge, MA: MIT Press.

Clark, H. (2009a). *Statement at the Annual Session of the Executive Board*, New York, May 26. http://content.undp.org/go/newsroom/2009/may/helen-clark-statement-at-the-annual-session-of-the-executive-board.en?categoryID=1684491&lang=en

Clark, H. (2009b). Statement on the Occasion of the Annual Session of the Executive Board of UNDP/UNFPA, May 26.

Clark, H. (2009c). *Statement by Helen Clark: The UN Conference on the World Financial and Economic Crisis*, New York, June 26. http://content.undp.org/go/newsroom/2009/june/helen-clark-un-conference-on-the-world-financial-and-economic-crisis.en?categoryID=1684491&lang=en

Clark, H. (2009d). Remarks by Helen Clark: Between Now and 2015: Moving the Development Agenda Forward, August 31, The Hague.

Clark, H. (2009e). *Remarks at the Pacific Leaders Event, UNFCCC Side-Event, Copenhagen*, December 14. http://content.undp.org/go/newsroom/2009/december/helen-clark-remarks-at-pacific-leaders-event-copenhagen.en?categoryID=1684491&lang=en

Clark, H. (2012a). *Statement of Helen Clark UNDP Administrator, Annual Meeting of the UNDP Executive Board*, Geneva, June 25. www.undp.org/content/undp/en/home/presscenter/speeches/2012/06/25/helen-clark-annual-meeting-of-the-undp-executive-board/

Clark, H. (2012b). *Why Tackling Climate Change Matters for Development*, Lecture by Helen Clark, UNDP Administrator and UNDG Chair, co-organized by the Stanford Woods Institute for the Environment & the Stanford Program in Human Biology, Stanford, CA, November 8. www.undp.org/content/undp/en/home/presscenter/speeches/2012/11/08/helen-clark-why-tackling-climate-change-matters-for-development-/

Cox, R. W. (1969). The Executive Head: An Essay on Leadership in International Organization, *International Organization* 23 (2): 205–230.

Derviş, K. (2006). *Statement to the Executive Board of UNDP/UNFPA*, January 24. http://content.undp.org/go/newsroom/2006/january/statement-Derviş-undp-unfpa-20060124.en?categoryID=349479&lang=en

Derviş, K. (2007a). *Speech at UNEP Governing Council Meeting*, February 6. content.undp.org/go/newsroom/2007/february/Derviş-unep-20070206.en?categoryID=1005974&lang=en a

---

[37] www.undp.org/content/undp/en/home/operations/leadership/administrator.html

Derviş, K. (2007b). *Kemal Derviş on Climate Change*, April 9. http://content.undp.org/go/newsroom/2007/april/kemal-Derviş-climate-change.en

Derviş, K. (2007c). *Speech at the Second Session of the Executive Board*, United Nations, New York, September 10.

Derviş, K. (2007d). *Op-Ed: Climate Change and Development: The Central Challenge of Our Time, UNDP,* United Nations, New York, November 28.

Executive Board of the UNDP/UNFPA (2004). *Report of the Executive Board on Its Work during 2004,* New York: Economic and Social Council.

Graham, E. (2015). Money and Multilateralism: How Funding Rules Constitute IO Governance, *International Theory* 7 (1): 162–194.

Hall, N. (2016a). The Institutionalisation of Climate Change in Global Politics. In G. Sosa-Nunez and E. Atkins (eds.), *Environment, Climate Change and International Relations*, E-International Relations, 60–75.

Hall, N. (2016b). *Displacement, Development, and Climate Change, International Organizations Moving beyond Their Mandates*, Abingdon: Routledge.

Hall, N. (2022), *Transnational Advocacy in the Digital Era*, Oxford: Oxford University Press.

Hall, N. and Persson, A. (2018). Global Adaptation Governance: Why Isn't It Legally Binding? *European Journal of International Relations* 24 (3): 540–566.

Hall, N. and Woods, N. (2018). Theorizing the Role of Executive Heads in International Organizations, *European Journal of International Relations*.

Hawkins, D., Lake, D., Nielson, D., and Tierney, M. (eds.) (2006). *Delegation and Agency in International Organizations*, Cambridge: Cambridge University Press.

Ivaanova, M. (2010). UNEP in Global Environmental Governance: Design, *Leadership, Location, Global Environmental Politics* 10 (1): 30–59.

Jinnah, S. (2014). *Post-Treaty Politics: Secretariat Influence in Global Environmental Governance*, Cambridge, MA: MIT Press.

Klingebiel, S. (1999). *Effectiveness and Reform of the United Nations Development Programme (UNDP)*, London: Frank Cass.

Malloch-Brown, M. (2011). *The Unfinished Global Revolution: The Road to International Cooperation*, London: Penguin Group.

Mearsheimer, J. (1995). The False Promise of International Institutions, *International Security* 19 (3): 5–49.

Melkert, A. (2008). Climate-Related Conflict and the Millennium Development Goals, Speech at the ECOSOC Luncheon on Climate-Related Conflict and the Millennium Development Goals, June 30. http://content.undp.org/go/newsroom/2008/june/amelkert-statement-environment-conflict-mdg.en?categoryID=1486741&lang=en

Melkert, A. (2009). *Speech to the UNDP/UNFPA Executive Board*, May 26. http://content.undp.org/go/newsroom/2009/june/ad-melkert-to-the-undp-unfpa-executive-board-.en?categoryID=1684491&lang=en

Mingst, K. and Karns, M. (2016). *The United Nations in the 21st Century*, Boulder, CO: Westview Press.

Murphy, C. N. (2006). *The United Nations Development Programme, a Better Way?* Cambridge, MA: Cambridge University Press.

Nielson, D. and Tierney, M. (2003). Delegation to International Organizations: Agency Theory and World Bank Environmental Reform, *International Organization* 57 (2): 241–276.

Park, S. and Weaver, C. (2012). The Anatomy of Autonomy. In Joel E. Oestrich (ed.), *International Organizations as Self-Directed Actors*, Abingdon: Routledge, 91–117.

Pfeffer, J. and Salancik, G. R. (1978). *The External Control of Organizations, A Resource Dependence Perspective*, Stanford: Stanford Business Classics.

Pollack, M. A. (2003). *The Engines of EU Integration: Delegation, Agency and Agenda Setting in the EU*, Oxford: Oxford University Press.

Schmid, L., Reitzenstein, A., and Hall, N. (2021). Blessing or a Curse? The Effects of Earmarked Funding in UNICEF and UNDP, *Global Governance: A Review of Multilateralism and International Organizations* 27 (3): 433–459.

Simons, M. (2013). To Ousted Boss, Arms Watchdog Was Seen as an Obstacle in Iraq, *New York Times*, October 13. www.nytimes.com/2013/10/14/world/to-ousted-boss-armswatchdog-was-seen-as-an-obstacle-in-iraq.html.

Stern, N. (2006). *The Stern Review on the Economics of Climate Change.* http://webarchive.nationalarchives.gov.uk/+/http:/www.hm-treasury.gov.uk/independent_reviews/stern_review_economics_climate_change/stern_review_report.cfm

Stokke, O. (2009). *The UN and Development*, Bloomington: Indiana University Press.

UNDP (2001). *Update on the UNDP Business Plans, Report to First Regular Session of the Board, DP/2001/CRP2*, New York: UNDP.

UNDP (2007). *Human Development Report 2007/2008: Fighting Climate Change: Human Solidarity in a Divided World.* New York: UNDP.

UNDP (2008a). *Climate Change Strategy*, New York: UNDP.

UNDP (2008b). *UNDP Strategic Plan 2008–2011, Accelerating Global Progress on Human Development, Executive Board of UNDP and UNFPA Annual Session, May 22*, New York: UNDP.

UNDP (2009). *Administrators Report to the Executive Board in UNDP Reports on Sessions*, New York: UNDP.

UNDP (2013). *Climate Change Adaptation and UNDP-GEF.* http://web.undp.org/gef/do_cc_adaptation.shtml

UNDP (2015). *Helen Clark: Speech on Climate Change and the Sustainable Development Goals*, December 7. www.undp.org/content/undp/en/home/presscenter/speeches/2015/12/07/climate-change-and-the-sustainable-development-goals.html

UNDP EEG (2013a). Climate Change Adaptation Bulletin, *A Quarterly Update of Activities* (13): 1.

UNDP EEG (2013b). Climate Change Adaptation Bulletin, *A Quarterly Update of Activities* (14): 1.

UNDP Evaluation Office (2008). *Evaluation of the Role and Contribution of UNDP in Environment and Energy*, New York: UNDP.

UNDP Evaluation Office (2010). *Evaluation of UNDP Contribution to Environmental Management for Poverty Reduction: The Poverty-Environment Nexus*, New York: UNDP.

UNDP Evaluation Office (2012). *Evaluation of UNDP Partnership with Global Funds and Philanthropic Foundations*, New York: UNDP.

UNDP Executive Board (2010). *Report of the Executive Board on Its Work during 2010*, New York: Economic and Social Council.

United Nations (2011). Rules of Procedure of the Executive Board of the United Nations Programme of the United Nations Population Fund and of the United Nations Office for Project Service, DP/2011/18, January.

Watkiss, P., Baarsch, F., Trabacchi, C., and Caravani, A. (2014). The Adaptation Funding Gap. In UNEP (ed.), *The Adaptation Gap*, Nairobi: United Nations Environment Programme (UNEP).

Woods, N., Kabra, S. S., Hall, N., et al. (2015). *Effective Leadership in International Organizations*, Geneva: Global Agenda Council on Institutional Governance Systems, World Economic Forum.

World Commission on Environment and Development (1987). *Report of the World Commission on Environment and Development: Our Common Future.* https://sustainabledevelopment.un.org/content/documents/5987our-common-future.pdf

Young, Z. (2002). *A New Green Order? The World Bank and the Politics of Global Environment Facility*, London: Pluto.

# 6

# Follow the Money

*Secretariat Financing as a Window on the Principal–Agent Relationship*

LYNN WAGNER AND PAMELA CHASEK

## 6.1 Introduction

If you attend a Conference of the Parties (COP) to a multilateral environmental agreement or the meetings of an intergovernmental science body, you will no doubt be caught up in the intrigue of the plenary debates and contact group discussions focused on substantive issues and national obligations to take action. Will the parties to the Convention on Biological Diversity (CBD) adopt a new global biodiversity framework when the ten-year agenda, as set out in the Aichi Targets, comes to a conclusion? Will the parties to the Paris Agreement on climate change under the United Nations Framework Convention on Climate Change (UNFCCC) finalize the rules on how countries can reduce their emissions using international carbon markets, as covered under Article 6? Will parties to the United Nations Convention to Combat Desertification (UNCCD) include land tenure as a new thematic area under the convention? Will the latest scientific assessment be adopted by the IPCC or the Intergovernmental Science-Policy Platform on Biodiversity and Ecosystem Services (IPBES) and clearly define human responsibility for causing and redressing global challenges?

While these headline agenda items will command most participants' attention, tucked away in parallel discussions, a small group of state delegates will be focused on the program and budget, with the aim of developing what will likely be the final set of decisions adopted at that meeting. The decisions of this group are essential to the operations of the convention or organization: Without an affirmative conclusion by this group, the lights will not remain on, the secretariat staff will not be paid, and the next meeting will not take place. In short, global cooperation through this forum cannot continue until this small group reaches agreement.

Member states to multilateral environmental agreements and intergovernmental science organizations establish secretariats to undertake a number of tasks required for their efficient operation. A central area of responsibility for secretariats is the organization of meetings of the COP or plenaries and other meetings of relevant subsidiary bodies, during which member states negotiate the ongoing work and focus of the treaty or organization, including the budget that funds the secretariat's activities over the subsequent year(s). A close examination of the decision-making process around these budgets offers a window into the principal–agent relationship between state parties and secretariats. State control of the purse strings is an important mechanism through which the principals in these intergovernmental organizations control the activities of their agents: Through their decisions on programs and budgets, states assert control over the focus of activity and level of ambition that secretariats can undertake. This chapter explores the dynamics and decisions taken regarding the secretariat budgets to shed light on this underexplored perspective in the principal–agent relationship. While the other contributions to this volume explore the ways that secretariats and international organizations can act independently of states, we explore one of the primary ways that states exercise control over secretariat activities.

We examine this relationship through case studies that consider budget-related decision-making processes and outcomes under the Rio Conventions – UN Framework Convention on Climate Change, Convention on Biological Diversity, and UN Convention to Combat Desertification – and two multilateral scientific bodies – the IPCC and the IPBES. The negotiations on the program and budget for the Rio Conventions reveal ways in which member states seek to control the activities of the secretariats through the budget structure. We also review the responses to budget crises by the secretariats of the scientific bodies and their members. The research draws on our participant observations of multiple multilateral environmental agreement (MEA) negotiations,[1] as well as the final decisions of the meetings we analyze. Before launching into the case studies, we begin the chapter with a review of the principal–agent literature as it applies to the cases we explore. The conclusion comments on what the cases suggest for the principal–agent relationship in multilateral environmental organizations.

## 6.2 Principals, Agents, and Resources

According to Biermann et al. (2009: 6), international bureaucracies are "agencies that have been set up by governments or other public actors with some degree of permanence and coherence and beyond formal direct control of single national governments … and that act in the international arena to pursue a policy." In other words, they are a hierarchically organized group of international civil servants

---

[1] The authors have been working as an executive editor (Chasek) and a writer (Wagner) for the International Institute for Sustainable Development's *Earth Negotiations Bulletin* since 1992 and 1994, respectively. In this capacity, they have attended COPs, observed budget contact group negotiations, and monitored decision-making for each Rio Convention COP.

with a given mandate, resources, identifiable boundaries, and a set of formal rules and procedures within the context of the establishing treaty, protocol, or charter. But what is "given" may be taken away, or at least restricted or redirected, albeit with a time lag built around annual or biennial decision-making at conferences of the principals.

The principal–agent focus is particularly useful for examining the relationship between member states and secretariats, as a special type of international organization that exists to administer a treaty or agreement. Principal–agent theory developed initially in the area of business studies focusing on the delegation processes within firms. It was later applied to US Congressional politics and European integration studies and has since been used in studies on international organizations (Bauer et al. 2009: 26–27; Elsig 2010). When applied to secretariats, principal–agent theory highlights the fundamental differences in the collective interests of national governments as the principals and the secretariats as the agents. It maintains that secretariats are able to develop autonomy from their principals and thus need to be understood as actors in their own right. In this perspective, secretariats can be seen as self-interested bodies that are predominantly interested in increasing their individual resources and competencies. Bauer et al. (2009: 27) indicate that the activities of secretariats need to be explained on the basis of their relationship to national governments that delegate authority to secretariats. Principal–agent theory can offer theoretical models to reveal the general influence of secretariats, as well as limits thereof, keeping in mind that the relationship between the principal and the agent is not fixed. The evolution of the relationship can be tracked by observing the program and budget negotiations.

The principal–agent concept is particularly on display when it comes to decisions on financing and budgets. Barnett and Finnemore (2004: 12) suggest that the study of international organizations as bureaucracies "puts the interactive relationship between states and IOs [international organizations] at the center of analysis" rather than assuming that states dictate to international organizations. But while their examination concludes that international organizations exercise behavioral autonomy from states, they recognize that states "provide the delegated authority and resources" for these organization, although "mechanisms of accountability have not kept pace with the power and reach of international organizations" (Barnett and Finnemore 2004: 170–171). The budget negotiations we focus on represent an accountability mechanism, albeit with a time delay, as they often take place on a two-year cycle.

An international public administration (IPA) focus, as presented in the introduction to this book, brings attention to the ways in which resources enter into the principal–agent relationship. This chapter considers the fourth of five sources of IPA influence, as identified by Bauer, Knill, and Eckhard (2017: 182–189).

Budgetary restrictions can be mechanisms of accountability through which principals limit or direct the activities of their agents. The examples of states establishing restrictions on how resources can be used, as presented in this chapter, reveal that this mechanism is as much a reaction to perceived overreaches by secretariats as it is a proactive set of guidelines for the principal's preferred direction.

In the conclusion to their study of secretariat influence, Biermann and Siebenhüner (2009: 330–333) distinguish among polity competence, resources, and embeddedness as some of the variables that help explain variation in the influence of international bureaucracies. They conclude that "there is no clear link between the availability of funds and the autonomous influence of bureaucracies" (Biermann and Siebenhüner 2009: 338), but this conclusion does not explore the give and take between the principal and agent in setting and resetting the availability of funds. We agree with Biermann and Siebenhüner (2009: 345) that "international bureaucracies are autonomous actors in world politics." Their principals' decisions on their programs and budgets would not be as belabored or respond to specific initiatives, as discussed later, if they were not. But while the accountability mechanism of the budget decisions cannot explain why one secretariat might be more ambitious (and influential) in its efforts to bring new activities into its program of activities than another, the possibilities for secretariat influence depend on its ability to mobilize resources for a particular activity. Biermann and Siebenhüner assign a lower importance to the polity – or legal, institutional, and organizational framework, including resources – than to the problem structure and the people and procedures of a given bureaucracy to explain variations in influence among secretariats. We suggest taking a closer look at the decisions taken around resources.

This chapter examines variables involved with the decision-making processes on resources as a mechanism of accountability and regulation of secretariat influence. The next section offers a short introduction to the funding sources and budgeting process for secretariats. It is followed by case studies related to program and budget decision-making under the Rio Conventions and the two multilateral scientific bodies.

## 6.3 Funding Avenues for Secretariats

In the UN system, funding has traditionally come from two sources: assessed and voluntary contributions. A system of assessed contributions requires member states to make financial contributions – or dues – as an obligation of membership. For example, the United Nations assesses mandatory contributions or dues to all members using the capacity-to-pay principle set out in the Charter of the United Nations, which takes into account the size of their economy (Graham 2015). The UN scale of assessments is modified by a ceiling and a floor placed on the proportion any single member state can pay to guard against tendencies by member states "to unduly minimize their contributions" or increase them unduly for prestige (UN General Assembly 1946, A/80). The United Nations General Assembly adjusts the scale of assessments every two years, and many UN specialized agencies and treaty bodies, including the Rio Conventions, use the United Nations General Assembly scale.

Voluntary contributions are usually considered to be extrabudgetary funds paid in order to finance specific operations or services (Francioni 2000). Unlike assessed funding, there is no legal obligation attached to voluntary funding systems (Archibald 2004). These systems lack the authority to allocate funding requirements across members, which leaves each member state with the ability to determine whether and how much to contribute. As a result, member state support for intergovernmental organizations funded by voluntary contributions can vary widely, with some gaining near universal support and some funded by a minority of members (Graham 2015). So while the relevant organization may adopt a budget every year or two, the actual funds received are determined by the individual donors. This creates a challenge for the secretariats that are often mandated by the member states to implement a work program but do not know from year to year whether they will have sufficient funds to do so and may have the added task of convincing individual member states or other donors to fund the voluntary portion of the budget. The biggest UN funds and programs – the United Nations Children's Fund, the United Nations Development Programme, the United Nations Population Fund, and the World Food Programme – are funded entirely by voluntary contributions.[2]

Further restricting the flexibility of secretariats is the fact that voluntary funds can often be "earmarked" for a particular purpose. Earmarked funding is provided by member states with conditions placed on the use of the funds. The practice of earmarking grew substantially in the 1990s, and by 2013, according to the Organisation for Economic Co-operation and Development (OECD 2015), the weight of funding to multilateral organizations that is earmarked for specific purposes, countries, or sectors represented 31 percent of total funding, with UN funds and programs receiving 76 percent of all funding as earmarked funds. A recent study finds the "growth in earmarked funding continues to outpace that in core funding" (Dag Hammarskjöld Foundation and UN Multi-Partner Trust Fund Office 2018: 10) and highlights that such funding is less flexible than core contributions, introducing questions for any inquiry of the ability of a secretariat to influence policy directions.

---

[2] Funding information for these funds and programs can be found at: UNICEF (www.unicef.org/partnerships/funding); UNFPA (www.unfpa.org/resources-and-funding); UNDP (www.undp.org/funding); and WFP (www.wfp.org/funding-and-donors).

Unlike the Rio Conventions, the IPCC and IPBES Secretariats are charged with producing scientific assessments on climate change and biodiversity/ecosystems services, respectively, and serve as intergovernmental science–policy interfaces. Also unlike the Rio Conventions, their budgets do not use the UN scale of assessments but rely entirely on voluntary contributions. The IPCC and IPBES procedures do not define any level of annual financial contribution each member state or observer organization must pay to support the budget and work program or the travel expenses for participants from developing countries and countries with economies in transition (IPCC 2017a).

With regard to private actors, Graham (2017) notes that as assessed contributions were supplemented by voluntary contributions, private actors also became eligible contributors. Like member states, private actors, including nongovernmental organizations, philanthropic organizations, and multinational corporations, often earmark their funding for specific purposes. For example, in 2015, specified voluntary contributions from foundations, corporations, and civil society to the UN system amounted to about USD 4 billion, or 14 percent of all specified voluntary contributions to the UN system (United Nations 2016). However, these trends primarily affect the UN development agencies (see Graham 2017; Seitz and Martens 2017).

As we explore in the next section, the process used to reach a decision on the amount of funding to be provided to an intergovernmental organization through assessed and voluntary sources is a function of the relationship between the principals (member states) and the agents (secretariats).

## 6.4 Push and Pull for Control in Programs and Budgets

The cases presented in this section explore the relationship between secretariats and parties from a number of angles. At each point, we find decisions made by the parties that directly or indirectly addressed or diminished the secretariat's initiatives.[3] We begin with an example that demonstrates a basic starting point in the principal–agent relationship: If the parties do not adopt a budget, the secretariat will cease to operate. This first case study also introduces a key focus of parties during budget negotiations: limiting the percentage increase in the budget rather than matching it to the level of programming required to achieve other decisions under negotiation at the same COP. This exploration provides background for reviewing the action and reaction from secretariats and parties in response to the

---

[3] "Parties" in this section refers to the collective will of all parties as reflected in COP decision documents. As might be expected, donor governments' preferences often prevail in the consensus-based budget discussions for the Rio Conventions.

increased level of programming secretariats have been assigned. When secretariats have presented parties with draft budgets that would significantly increase their funding, parties have responded by adopting guidelines for future budget proposals that restrict the percentage increase those future budgets can incorporate. We then review the level of assessed and supplementary budget components over time for the three Rio Conventions, noting that the former has been consistent within each convention and across the three conventions and the latter has been the source of fluctuation. Finally, we present the experience of IPCC and IPBES in the face of budget shortfalls, to which the parties ultimately responded with funding rather than cede control based on the requirements of unconventional funders.

## *Parties Control the Switch to Keep the Lights On*

At its most basic, the continued operation of the secretariat is on the line with each budget negotiation. A COP may decide to push a decision on reducing emissions or cooperating on biosafety issues to the next meeting if the parties cannot reach an agreement. But if the budget is not adopted, the organizing entity for that next meeting – the secretariat – will not be able to operate. Without a budget, funds will not be allocated for secretariat staff salaries, office requirements, and preparations for the next meeting of the COP. This point was illustrated during the negotiations for the eighth session of the UNCCD COP, which took place in September 2007 in Madrid, Spain.

During this UNCCD COP, a Japanese delegate had consistently, but not forcefully, voiced his country's position that the overall budget should be the same as for the previous biennium: zero nominal growth. The program and budget contact group was meeting in parallel to the negotiations on the new strategy for the convention, which the parties had called for to help define the convention's purpose and guide its approach to combatting drought, land degradation, and desertification. In addition, the convention had just undergone a change in leadership. The first executive secretary had had a combative relationship with developed country parties over the role of the secretariat in implementation activities, which had manifested itself in budget decisions that sought to control the secretariat's scope (Wagner and Mwangi 2010). Despite the Japanese delegate's position, the draft budget decision that was sent to the closing plenary in Madrid provided for a 5 percent increase in the euro value of the budget, with clear secretariat support. However, the Japanese delegate had only agreed to the proposal *ad referendum* in a contact group.[4] While many delegates left the conference center because they expected the final adoption of decisions to be without incident, the Japanese delegate contacted his capitol and

---

[4] Under the condition that the agreement would be confirmed by his state.

was instructed not to accept the draft budget (Conliffe et al. 2007). General chaos ensued through an all-night scramble to determine what would happen next.

The solution was to hold an extraordinary COP before the end of the year, at UN headquarters in New York, to adopt a budget. However, the negotiations continued along the same lines during that one-day event, with Japan holding to its position of zero nominal growth. It became evident that this country desired to set a precedent for other MEA budget negotiations that year. Ultimately, UNCCD delegates adopted a budget with a 4 percent increase, although 1.2 percent of it (EUR 185,000) was to be met, "without creating a precedent for this or any other convention," by the government of Spain (which held the COP presidency) as a way to break the deadlock. With this compromise, the negotiations concluded at 4:00 a.m., and the secretariat's lights remained on for two more years (Chasek 2007: 2).

The UNCCD COP8 budget negotiations illustrate the principal's ultimate authority over maintaining a functioning agent. While these talks were held up by one party, the consensus required for all Rio Convention outcomes could be similarly impeded by any number of parties. The reactions of MEAs to the restrictions on global meetings due to the global COVID-19 pandemic reinforce this point. While the pandemic resulted in the postponement of many COPs, parties convened extraordinary COPs using the "silence procedure" to adopt programs and budgets in order to keep the secretariats functioning until global meetings could resume (see, e.g., Sollberger 2020).

### *Reigning in Secretariat Budget Proposals*

The focus of parties on limiting the growth of the core budget, regardless of the level of ambition in the substantive expectations for the convention's program for the biennium, can be further illustrated by the experience of the 2009 UNCCD COP, which took place in Buenos Aires two years after the protracted budget talks in Madrid. This COP followed the adoption of this MEA's new ten-year strategy. Despite the fact that these talks also came on the heels of the financial crash, the executive secretary attempted to set the tone for the budget discussions by presenting a proposed budget with a 16 percent increase over the previous biennium. Negotiators who had come into the talks with instructions to hold the growth of the budget to a much lower percentage were not prepared to engage in a discussion of this proposal, and were even concerned with whether the executive secretary was in touch with the political environment in which he needed to operate. Negotiations focused on three options to increase the budget (5 percent, 4.29 percent, and 3.36 percent), none of which was close to the secretariat's proposal. Negotiators eventually settled on the middle option (Aguilar et al. 2009).

Negotiations on the budget are often hampered by differences in participants' approaches to framing the proposed budget increase. Resulting tensions may exist between developing and developed country parties as well as between the member states and the secretariat. As noted, many of the parties will enter the budget negotiations with instructions from their government regarding an acceptable percentage increase (or lack thereof) over the budget adopted for the prior year or biennium. At the same time, the negotiators are facing the challenge that parallel negotiations regarding the programs and projects that the secretariat will be asked to implement are taking place, and the budget negotiations should provide the resources for those programs and projects. These competing priorities and influences on budget negotiations can lead to a disconnect between the ultimate decision on the budget and the substantive decisions adopted by the COP. Behind the scenes at the third CBD COP, for example, the executive secretary developed a tally of the estimated cost for each decision as it was adopted, but rumor has it that he decided not to share the information with delegates because the tally had far overtaken the budget level under discussion in the program and budget contact group (Carpenter et al. 1996).

The secretariat, although officially not a party to the negotiations, often leads the messaging about the need to connect the budget with the ambition identified in the substantive decisions, beginning with the background documentation prepared for the COP. The secretariat tables what serves as a starting point for the budget negotiations in its background documentation provided to the parties. As in any negotiation, this proposal can frame the negotiations and influence the ultimate size of the budget. The secretariat also faces the possibility of a backlash from delegates if the proposal is deemed to be unreasonable. If secretariats have a free hand in crafting this budget, these agents could frame the principals' debate over the budget level. But the parties have taken steps to curtail this potential area of secretariat influence.

The UNCCD executive secretary's strategy at COP9 was particularly questionable given that the previous COP had collapsed in the final hours due to the size of the budget. While the strategy did not seem to take the previous budget negotiation process into account, the parties reacted to the secretariat's perceived overreach by exerting control over future budget proposals. In addition to adopting a budget that was very different from the size proposed by the UNCCD Secretariat, the UNCCD parties at the 2009 COP took a step to take control of the framing of future budget negotiations by placing explicit instructions in the program and budget decision regarding the budget proposals that the secretariat should include in its documentation for the next COP. UNCCD COP9 included the instruction for the secretariat to include budget scenarios reflecting zero nominal growth and zero real growth in the documentation for the next COP (Aguilar et al. 2009).

Similar decisions have been taken by the parties to other conventions. For example, the parties to the CBD began requesting specific budget proposals from the CBD executive secretary at COP9, with decision IX/34 requesting the executive secretary to provide three alternatives for the budget.[5] These alternatives were to include one option based on an assessment of the required rate of growth for the program budget, one option that would maintain the program budget in real terms, and one option that would maintain the program budget in nominal terms. At CBD COP13, the parties reduced the executive secretary's freedom in assessing the required rate of growth, specifying that it should not exceed 5 percent above the previous biennium in nominal terms.

These decisions have been taken by the parties (principals) to reign in secretariats' (agents) ambition to frame the program and budget negotiations. In response to secretariats' efforts to match the proposed budget with the substantive level of activity that the parties' substantive decisions suggest is necessary, parties have taken steps to frame the discussion as a matter of inflation or limited growth. These examples also demonstrate the contentious nature of the program and budget discussions, with the secretariat pushing for higher levels and the parties focused on limiting the level of growth, often based on percentage amounts rather than the program levels adopted in other decisions.

The parties have been fairly consistent in holding the growth of the core budget and have also taken steps to control the framing of the budget negotiation by instructing the secretariat about the proposals that can be submitted to the COP. The greatest room for variation in funding levels, and possibly for secretariats to access funding for the issues they have introduced, would be through voluntary funding, to which we now turn.

### *Assessed versus Supplementary Budgets: Space for Ambition?*

The structure of the budgets for the three Rio Conventions clearly incorporates a division between assessed and voluntary funding and is central to the examination of the relationship between secretariats and member states. Beginning with the first COP for each Rio Convention, the parties have adopted a single operational budget for each convention (referred to here as the core budget). The core budget is based on assessed contributions, using the UN scale of assessments to determine each party's contribution. Each CBD budget even specifies that the total of this core budget is the "budget to be shared by parties."[6] A close examination of the level

---

[5] "Integrated Programme of Work and Budget for the Convention and Its Protocols," Decision COP XIV/37, www.cbd.int/decisions/cop/14/37

[6] See for example the final line (15) on page 126 of the budget adopted by CBD COP3 (CBD 1997).

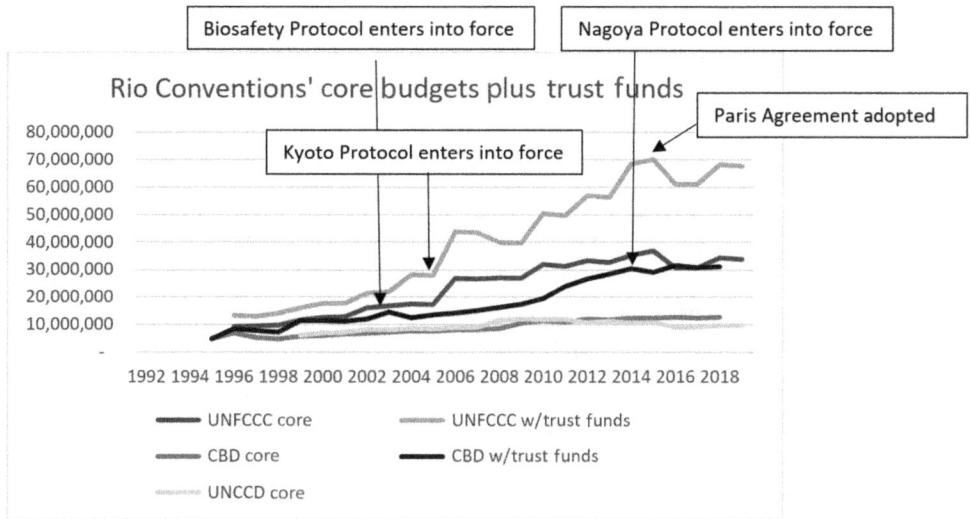

Figure 6.1 Rio Conventions' core budgets plus trust funds
*Source*: UNFCCC Reports of the Conference of the Parties (available at https://unfccc.int/documents); CBD reports on the Administration of the Convention and the budget for the trust funds of the Convention (available at www.cbd.int/decisions/); UNCCD Reports of the Conference of the Parties (available at www.unccd.int/convention/official-documents).

of assessed funding reveals that it has remained fairly constant throughout the Rio Conventions' history and is generally aimed at supporting the basic framework of the MEA that the parties established (see Figure 6.1). Earmarked and voluntary funds are directed to the funders' preferred activities and may therefore focus on projects identified by an entrepreneurial secretariat. However, even these voluntary funds are not always easy to use, even if secured, as parties have instituted constraints on both assessed and voluntary budgets in some budget decisions, often in response to secretariat-initiated efforts. We begin this examination with a focus on parties' approach to setting the core budget levels for the three Rio Conventions, followed by an overview of the differences between the core and supplementary budgets for the conventions.

As presented in Figure 6.1, the core budgets of the three Rio Conventions were relatively constant over their first twenty-plus years. Furthermore, the size of the core budgets for the Rio Conventions have been closely and significantly correlated with one another over time (UNFCCC and CBD = .942; UNFCCC and UNCCD = .795; CBD and UNCCD = .714).[7]

---

[7] All correlations are bivariate Pearson correlations and are significant at the 0.01 level (two-tailed). They measure the strength and direction of linear relationships from the annual budgets for the Rio Conventions from inception of each convention to the most recently adopted budget.

Figure 6.1 also illustrates that the adoption of protocols shows up in the supplementary budgets, not the core budget, and does not necessarily result in a lasting increase in the budget. After an initial period at higher levels following the entry into force of the CBD Biosafety and Nagoya Protocols, the CBD budget decreased slightly. The entry into force of the Kyoto Protocol resulted in a lasting higher level of funding for the UNFCCC supplementary budget, although the peak level reached immediately after entry into force was not maintained. The parties have essentially set funding at a maintenance level for the Rio Convention secretariats through the core budgets. Additional activities have required each secretariat to secure specific funding, which implies that the secretariat secured the approval of the funding party but not necessarily the entire COP through negotiated supplementary budget agreements.

A close look at the budgeting structure under the UNFCCC and CBD reveals that parties have sought to exert control by establishing trust funds with specific purposes, although these trust funds still provide vehicles for voluntary and variable funding for new initiatives. For example, with each budget cycle,[8] the UNFCCC parties have adopted a core budget as well as budgets for the Trust Fund for Participation and the Trust Fund for Supplementary Activities.[9] The UNFCCC core budget has tripled in size over its first twenty-three years, growing from USD 9,229,700 in 1996 to USD 33,840,957 for 2019.[10] Meanwhile, the total budgets (core plus trust funds) have grown over five times as large, from USD 13,311,150 in 1996 to USD 67,659,810 for 2019. The funding for ensuring wide participation in the work of the UNFCCC Secretariat dropped between 1996 and 2019, while the funding for supplementary activities has grown. In 1996, the specified cap for the Trust Fund for Participation (USD 2,770,990) exceeded the cap for the Trust Fund for Supplementary Activities (USD 1,310,460). By 2019, although ensuring that all parties are able to participate in the meetings of the COP remains important, the funding for supplementary activities (USD 32,090,651) far exceeded the lowest option listed for the funding for participation (USD 1,728,202).

Among the many activities included in the UNFCCC, the supplementary activities fund has been funding for the Momentum for Change initiative. This example offers an interesting case for how supplementary funding and secretariat initiative can intersect. While this funding is included in the 2016–2017 and 2018–2019

---

[8] The UNFCCC program and budget is adopted every two years, even though the COP meets every year. The budgets for the CBD and UNCCD are also based on negotiations every biennium; these latter two conventions have moved to having COPs every two years.

[9] In recent years, the UNFCCC Trust Fund for Participation has indicated a range of funding that could be collected to provide funding for developing country representatives to attend meetings organized under this convention. Calculations for the UNFCCC have used the lower end of this proposed range.

[10] Sources are various years of the UNFCCC Reports of the Conference of the Parties (available at https://unfccc.int/documents). In 2007, the budget switched from being denominated in US dollars to being denominated in euros. The amounts reported here are calculated using the UN operational rates of exchange (https://treasury.un.org/operationalrates/OperationalRates.php).

program budgets, the initiative itself began in 2011 at the initiative of the executive secretary and with funding from several foundations (UNFCCC 2014). The incorporation of this initiative into the supplemental budget means that the parties recognized the value of the project, but it also brings at least a portion of the budget under party constraints going forward.

Unlike the UNFCCC budget, the CBD parties added the Proposed Budget Covered by Voluntary Contributions (equivalent to the UNFCCC's Trust Fund for Supplementary Activities) during the CBD's second budget year and the Participation Trust Fund (equivalent to the UNFCCC's Trust Fund for Participation) during the third budget year. A third trust fund – the Participation Trust Fund for Indigenous Peoples – was added during the CBD's fifteenth budget year. The establishment of the latter trust fund in itself demonstrates how the principals have exerted control over the agents. The participation funds for indigenous peoples could have been comingled with the existing participation trust fund, but the parties wanted a full accounting for the clearly specified funding purpose. The CBD core budget in 1995 was USD 4,787,000.[11] It had more than doubled by 2018, growing to USD 12,706,200. By contrast, the total budget was six times as large, growing from USD 4,787,000 in 1995 to USD 31,187,350 in 2018.

As the previous review of how parties frame the budget negotiation suggests, growth in the assessed budgets for the Rio Conventions has been relatively restricted and limited. With the trust funds for participation, parties have funneled funding to principal-endorsed activities. In the case of the CBD, even the background of the participants has been specified, adding further party control to the use of the funds. The supplemental activities trust funds offer the greatest room for new activities and initiatives. This funding source is where we see additional funds coming into secretariats with the addition of new protocols. The supplementary trust funds have also provided a vehicle for moving some initiatives under a party-funded umbrella, as was the case for the Momentum for Change initiative. The next section explores two cases in which secretariats flirted with securing outside funding for unfunded activities, only to have the parties step up their funding commitments in recognition that such funding would reduce their control.

### *Filling Budget Shortfalls*

Because international institutions are vulnerable to budgetary instability, they may need to seek to mobilize "budgetary means from alternative sources in order to reduce their dependence on member state contributions" (Bauer, Knill, and

---

[11] Sources are various years of the Administration of the Convention and the budget for the Trust Funds of the Convention (available at www.cbd.int/decisions/).

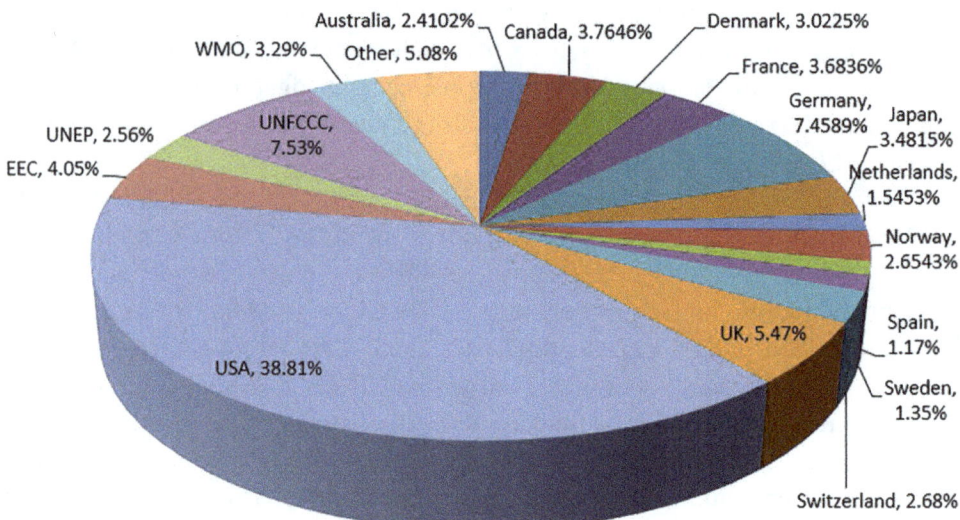

Figure 6.2  Contributors to the IPCC Trust Fund: 1989–2016
*Source*: ipcc.org (financial reports for each session of the IPCC)

Eckhard 2017: 187; see also Patz and Goetz 2017). Our final case studies examine this reaction to instability, in the form of budget shortfalls in two scientific multilateral bodies. These cases provide insights into instances in which secretariats solicited extrabudgetary funding and the response this effort prompted on the part of the parties. In these cases, the principals recognized that their influence would diminish if they were not providing the funding. To understand the shortfalls and options for solutions, we first need to understand these bodies' funding sources.

Since its inception in 1988 through 2017, fifty-four governments and organizations have contributed CHF 119,531,971 to the IPCC Trust Fund (IPCC 2017a). Of these, seventeen governments and organizations have contributed 95 percent of the funds: Australia, Canada, Denmark, European Commission, France, Germany, Japan, Netherlands, Norway, Spain, Sweden, Switzerland, the United Kingdom, the United States, the United Nations Environment Programme (UNEP), UNFCCC, and the World Meteorological Organization. The United States alone, until 2017, had contributed nearly 39 percent of the funds (see Figures 6.2 and 6.3). These figures do not include in-kind contributions, such as support for the IPCC Technical Support Units, publications, translation, meetings, and workshops. In addition to relying on a comparatively small donor base (16 percent of member states), funding varies from year to year. Some funders have contributed only sporadically. Others change the amount they give from year to year – either due to fluctuating exchange rates or their own changing budget priorities (see IPCC 2017b, Annex I, for a complete list).

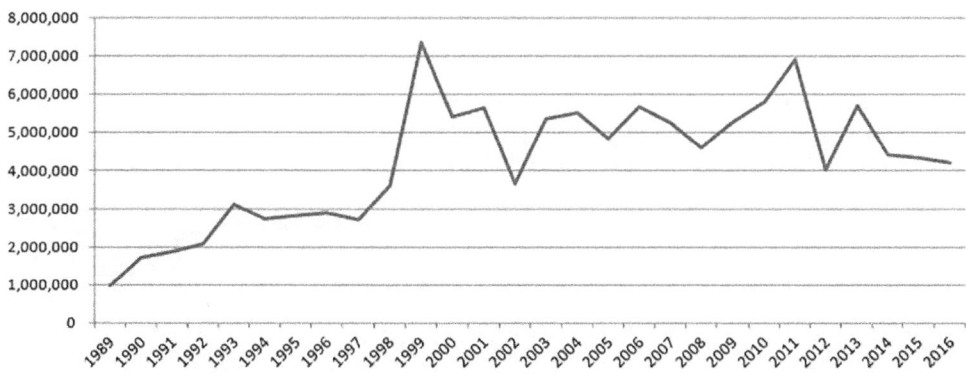

Figure 6.3 IPCC Trust Fund contributions: 1989–2016
*Source*: ipcc.org (financial reports for each session of the IPCC)

IPBES, which was established in 2012, relies on three types of resources: cash contributions to the trust fund; in-kind contributions to support the implementation of the work program; and the leveraging activities of its partners (IPBES 2017a). According to the IPBES Financial Procedures (IPBES 2015), the trust fund is open to voluntary contributions from all sources, including governments, UN bodies, the Global Environment Facility, other intergovernmental organizations, and other stakeholders, such as the private sector and foundations, although the amount of contributions from private sources must not exceed the amount of contributions from public sources in any biennium. The Financial Procedures note that financial or in-kind contributions from governments, the scientific community, other knowledge-holders, and stakeholders will not orient the work of the platform, maintaining the member states as the principals.

As of December 31, 2017, 22 out of 127 member states contributed USD 31,141,874 to the IPBES Trust Fund (IPBES 2017b) (see Figure 6.4). Of these, four governments contributed 77 percent of the funds: Germany, Norway, the United Kingdom, and the United States. Norway and Germany alone have contributed 58 percent. Most of the donors are OECD countries and no international organizations contributed to the trust fund. The amount of contributions to the trust fund generally ranges between USD 3.1 million and USD 4.2 million, with the exception of 2014, which benefited from a USD 8.1 million contribution from Norway at the start of the first work program (see Figure 6.5). The number of donors each year has ranged from thirteen to seventeen, with the exception of 2012, the year IPBES was established. Cash contributions came exclusively from governments. Some donor governments contributed on a regular basis, while others did not, and the amount of each contribution varied (IPBES 2017a).

In-kind contributions amounted to an additional USD 2,819,643 from fifteen governments, four intergovernmental organizations, two universities, a graphic

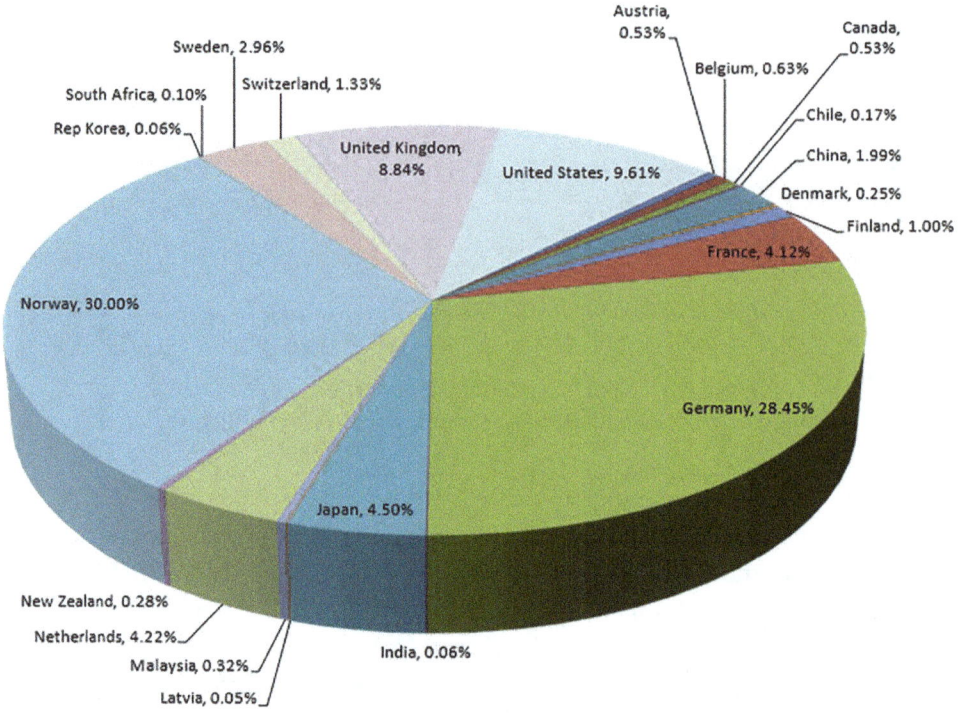

Figure 6.4   Contributors to the IPBES Trust Fund: 2012–2017
*Source*: IPBES (2017b)

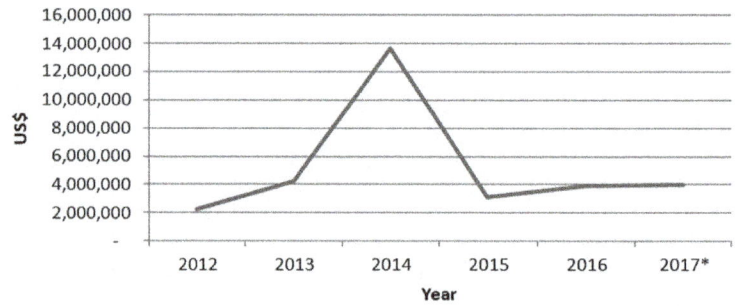

Figure 6.5   IPBES Trust Fund contributions: 2012–2017
* Does not include pledged amounts not received as of December 31, 2017.
*Source*: IPBES (2017b)

design company, and an individual. In-kind contributions are defined as direct support, not received by the trust fund, for activities either scheduled as part of the work program, which otherwise would have to be covered by the trust fund, or organized in support of the work program. In-kind contributions cover a wide range of activities, including: (i) provision of time and expertise at no cost to IPBES by the experts that are members of assessment and other expert groups – an in-kind contribution without which the implementation of the work program of IPBES would not be viable; (ii) costs of participation in IPBES meetings by experts from developed countries that are not eligible for financial support; (iii) provision of technical support for specific deliverables by institutions hosting technical support units; (iv) provision of meeting facilities and logistical support for specific meetings; and (v) provision of data such as data relevant to indicators, access to knowledge otherwise available only for a fee, or free access to existing digital infrastructure (IPBES 2017a).

The IPCC and IPBES have both struggled with funding shortfalls due to the voluntary nature of contributions. Since the early 1990s, the IPCC has sought ways to regularize the budget, increase the donor base, and share the costs more broadly among member states. Year after year, the IPCC Secretariat has sent letters to member states and organizations requesting contributions. Yet, despite best efforts, the funding base did not grow and the contributions continued to vary each year. Substantial contributions by a few member states in the 1990s and early 2000s allowed the IPCC to constitute cash reserves, as expenditures were far below the level of contributions. More recently, however, the reduced contributions as well as number of contributors have decreased the IPCC cash reserves, especially as the level of expenditures has been higher than the income received (IPCC 2017a). As a result, the reserves decreased from CHF 13.4 million in 2010 to CHF 5.8 million in January 2017. While there is no specific requirement as to the size of the reserves in the IPCC Trust Fund, the financial rules provide that a working capital reserve shall be maintained to ensure continuity of operations in the event of a temporary shortfall of cash (IPCC 2017a).

Concern about the ability of the IPCC to complete this assessment cycle led IPCC-45 in Guadalajara, Mexico (March 2017), to establish an *Ad Hoc* Task Group on Financial Stability (ATG-Finance) with the purpose of exploring avenues for financial stability of the IPCC, including funding options. Also at IPCC-46, the IPCC Financial Task Team reported that the approved IPCC Trust Fund budget – the IPCC fundraising target for 2017 – was CHF 8.3 million. As of January 1, 2017, the opening cash balance in the IPCC Trust Fund was CHF 5.8 million. By June 29, 2017, the total amount of voluntary contributions received equaled only CHF 992,670. A projected funding gap of CHF 5.7 million would exhaust the cash reserves of the IPCC Trust Fund (IPCC 2017a). Hence, there was concern that without more funding the IPCC would not be able to implement its work program which included the special report on 1.5°C and the seventh assessment cycle products.

At roughly the same time, IPBES found itself in a similar financial shortfall. At the fifth meeting of the IPBES Plenary in March 2017, IPBES Executive Secretary

Anne Larigauderie presented the budget and draft fundraising strategy. She highlighted that a realistic estimate of current IPBES activities, without launching new assessments and assuming a regular level of national contributions, would require an additional USD 3.4 million for 2017–2019 to complete ongoing activities. The meeting was dominated by discussions on the budget and resulting tensions regarding whether three pending assessments in the platform's first work program could be initiated and in what order they should be initiated if funds were insufficient for all three. Delegates ultimately adopted a budget that did not allow for the initiation of any pending assessments to reduce the risk of incurring a budget shortfall in 2018 and allowed for the secretariat to proceed in "survival mode" (Jungcurt et al. 2017).

In light of funding shortfalls and reduced budgets, the secretariats for both the IPCC and IPBES and some member states looked for alternative sources of funding. The IPCC and IPBES secretariats, for example, considered options for increasing the contributions from governments, including assessed contributions, in-kind contributions, and broadening the donor base in terms of contributing governments; exploring means to mobilize additional resources, including from UN organizations and others (e.g., UNEP, Global Environment Facility, Green Climate Fund) and evaluating their potential implications, in particular issues related to conflict of interest and legal matters; and providing guidance on the eligibility of potential donors, in particular the private sector. They also explored the viability of contributions from science/research and philanthropic institutions and the option for crowd funding (IPBES 2017a; IPCC 2017a).

Yet when these options were discussed by the IPCC and IPBES plenaries, member states expressed concern that expanding the sources of funding could have repercussions and could decrease their influence on these inter*governmental* organizations. For example, philanthropic foundations, in particular, can have enormous influence on political decision-making and agenda-setting in international organizations. This is most obvious, according to Seitz and Martens (2017), in the case of the Gates Foundation, which exerts influence on the United Nations not only through their direct grant-making but also through the placement of foundation staff in decision-making bodies of international organizations, including the World Health Organization (WHO). It further uses matching funds to influence governments' funding decisions and, thus, priority-setting in the WHO. Similarly, some member states expressed concern about private sector funding that could run the risk of conflict of interest and could damage the panels' integrity and independence (Mead et al. 2017).

The presentation of these types of funding options to member states led to a clear response. In addition to expressing concerns about conflicts of interest, some member states worried that if the secretariat received funding from a greater number of nongovernmental entities, the principal–agent relationship could erode and the secretariat could become more autonomous from the member states. In response to these concerns, the member states of both panels increased their voluntary contributions. In the second half of 2017, for example, IPCC contributions nearly doubled that of the first half of the year (IPCC 2018). IPBES voluntary contributions from member states also improved (IPBES 2017b).

## 6.5 Conclusion

The examples reviewed in this chapter demonstrate ways in which program and budget negotiations provide a mechanism through which the principals and agents act and react to influence the direction the institution will take. At the base of the relationship between the parties and secretariat is the fact that principals' affirmative decision is required at regular intervals to adopt a basic budget and keep the agent up and running. The level of funding for the basic core level of activities has been closely regulated by the principal, with decisions even instructing the secretariat regarding the size of the budget growth that can be proposed for subsequent budgets. This mechanism means principal control in reaction to the agents' initiatives comes with a time lag, but it also demonstrates the premise of this volume – the principals' reaction means that agents are seeking ways to exert their influence in the first place.

Additional trust funds in the Rio Conventions have provided room for funding a variety of initiatives, although the parties have maintained a level of control over these funds as well. While provisions are made for secretariats to solicit and secure additional, voluntary, funding, parties have set limits on this funding and have even stepped in when they recognized their control may be impacted if they were not supplying the core budgetary provisions.

Budgetary restrictions can be mechanisms of accountability through which principals limit or direct the activities of their agents. Principal–agent framing for examining the interactions between member states and secretariats in program and budget decision-making reveals ways in which these decisions are used as a mechanism for accountability. This check on the alignment of member state and secretariat priorities and directions is also a function of the structure of the budget, with much of the funding for implementation activities being specifically delineated. The designation of how much the secretariat can allocate from voluntary funds to specific activities – some of which were introduced due to secretariat initiative – adds a layer of control for the principal. The recognition by the principal that it may lose a level of control if it allows the secretariat to solicit outside funds from nongovernmental entities further illustrates how this mechanism plays a role in limiting secretariat influence.

Governments recognize that they have greater control over the organization, its activities, and the budgeting process when they control the budget – either through their contributions or by withholding those contributions. Funding rules specify whether the collective principal holds the primary mechanism of influence and control – the power of the purse – or whether that source of influence and accountability will sit with individual nongovernmental donors (Graham 2015). As the member states of the IPCC and IPBES realized, if decisions over funding shift from government donors to private sector donors or other nongovernmental donors, the influence of governments will likely decrease over time and the principal–agent relationship as well as the intergovernmental nature of the organization could come into question. While a decision by the parties to incorporate an activity that began as a secretariat initiative – such as the UNFCCC's Momentum for Change initiative – could be seen as a sign of secretariat influence over the agenda, it also serves as a mechanism through which the parties can reassert control over the agenda.

Through their decisions on programs and budgets, states continue to assert control over the focus of activity and level of ambition that secretariats can undertake. As these case examples illustrate, many governments are holding onto their position as principals in intergovernmental environmental organizations in order to hold the agent (the secretariats) accountable and regulate secretariat influence while limiting the influence of nongovernmental actors, including the private and civil society sectors. By continuing to wield the power of the purse, governments (principals) will continue to keep these organizations, especially the secretariats, under their control.

## References

Aguilar, S., Conliffe, A., Russo, L., Wagner, L., and Xia, K. (2009). Summary of the Ninth Conference of the Parties to the UN Convention to Combat Desertification: 21 September–2 October 2009, *Earth Negotiations Bulletin* 4: 229.

Archibald, J. (2004). Pledges of Voluntary Contributions to the United Nations by Member States: Establishing and Enforcing Legal Obligations, *George Washington International Law Review* 36 (2): 317–376.

Barnett, M. and Finnemore, M. (2004). *Rules for the World: International Organizations in Global Politics*, Ithaca, NY: Cornell University Press.

Bauer, M. W., Knill, C., and Eckhard, S. (2017). International Public Administration: A New Type of Bureaucracy? Lessons and Challenges for Public Administration Research. In M. W. Bauer, C. Knill, and S. Eckhard (eds.), *International Bureaucracy: Challenges and Lessons for Public Administration Research*, Basingstoke: Palgrave Macmillan, 179–198.

Bauer, S., Biermann, F., Dingwerth, K., and Siebenhüner, B. (2009). Understanding International Bureaucracies: Taking Stock. In F. Biermann and B. Siebenhüner (eds.), *Managers of Global Change: The Influence of International Environmental Bureaucracies*, Cambridge, MA: MIT Press, 15–36.

Biermann, F. and Siebenhüner, B. (2009). The Influence of International Bureaucracies in World Politics: Findings from the MANUS Research Program. In F. Biermann and

B. Siebenhüner (eds.), *Managers of Global Change: The Influence of International Environmental Bureaucracies*, Cambridge, MA: MIT Press, 319–349.

Biermann, F., Siebenhüner, B., Steffen, B., et al. (2009). Studying the Influence of International Bureaucracies: A Conceptual Framework. In F. Biermann and B. Siebenhüner (eds.), *Managers of Global Change: The Influence of International Environmental Bureaucracies*, Cambridge, MA: MIT Press, 37–74.

Carpenter, C., Chasek, P., Gardner, E., et al. (1996). Third Session of the Conference of the Parties to the Convention on Biological Diversity: 4–15 November 1996, *Earth Negotiations Bulletin* 9: 65.

Chasek, P. (2007). Summary of the First Extraordinary Session of the Conference of the Parties to the UNCCD: 26 November 2007, *Earth Negotiations Bulletin* 4: 207.

Conliffe, A., Mwangi, W., Wagner, L., and Xia, K. (2007). Summary of the Eighth Conference of the Parties to the Convention to Combat Desertification: 3–14 September 2007, *Earth Negotiations Bulletin* 4: 206.

Convention on Biological Diversity (CBD) (1997). *Budget of the Trust Fund for the Convention on Biological Diversity*. www.cbd.int/doc/decisions/cop-03/full/cop-03-dec-en.pdf

Dag Hammarskjöld Foundation and UN Multi-Partner Trust Fund Office (2018). *Financing the UN Development System: Opening Doors*. Uppsala, Sweden: Dag Hammarskjöld Foundation. www.daghammarskjold.se/wp-content/uploads/2018/09/financial-instr-report-2018-interactive-pdf_pj.pdf

Elsig, M. (2010). Principal–Agent Theory and the World Trade Organization: Complex Agency and "Missing Delegation," *European Journal of International Relations* 17 (3): 495–517.

Francioni, F. (2000). Multilateralism à la Carte: The Limits to Unilateral Withholdings of Assessed Contributions to the UN Budget, *European Journal of International Law* 11 (1): 43–59.

Graham, E. R. (2015). Money and Multilateralism: How Funding Rules Constitute IO Governance, *International Theory* 7 (1): 162–194. DOI: 10.1017/S1752971914000414.

Graham, E. R. (2017). Follow the Money: How Trends in Financing Are Changing Governance at International Organizations, *Global Policy* 8 (S5): 15–25. DOI: 10.1111/1758-5899.12450.

IPBES (2015). *Financial Procedures for the Intergovernmental Science-Policy Platform on Biodiversity and Ecosystem Services*. www.ipbes.net/system/tdf/downloads/IPBES_financial_procedures.pdf?file=1&type=node&id=15253

IPBES (2017a). *Draft Fundraising Strategy*. Annex II to Decision IPBES-5/6 in Report of the Plenary of the Intergovernmental Science-Policy Platform on Biodiversity and Ecosystem Services on the Work of Its Fifth Session. IPBES/5/15, April 11. www.ipbes.net/system/tdf/ipbes-5-15_en.pdf?file=1&type=node&id=15537

IPBES (2017b). *Financial and Budgetary Arrangements for the Platform*. IPBES 6/9, December 11. www.ipbes.net/system/tdf/ipbes-6-9_financial_and_budgetary_arrangements_for_the_platform.pdf?file=1&type=node&id=16560

IPCC (2017a). *Ad Hoc Task Group on Financial Stability of the IPCC Report on the Financial Stability of the IPCC*. IPCC-XLVI/Doc. 8. www.ipcc.ch/apps/eventmanager/documents/47/150820170305-Doc.%208%20-%20Report%20on%20the%20Financial%20Stability%20of%20the%20IPCC.pdf

IPCC (2017b). *IPCC Trust Fund and Budget*. IPCC-XLVI/Doc. 2. www.ipcc.ch/apps/eventmanager/documents/47/040720170428-Doc.%202Budget.pdf

IPCC (2018). *IPCC Trust Fund Programme and Budget*. IPCC-XLVII/Doc. 2, Rev. 1. www.ipcc.ch/site/assets/uploads/2018/04/120320180135-Doc.-2-Rev.-1Budget.pdf

Jungcurt, S., Antonich, B., Louw, K., and Recio, E. (2017). Summary of Stakeholder Day and the Fifth Session of the Plenary of the Intergovernmental Science-Policy Platform on Biodiversity and Ecosystem Services: 6–10 March 2017, *Earth Negotiations Bulletin* 31: 34. http://enb.iisd.org/download/pdf/enb3134e.pdf

Mead, L., Cardenes, I., Gutiérrez, M., and Woods, B. (2017). Summary of the 46th Session of the Intergovernmental Panel on Climate Change: 6–10 September 2017, *Earth Negotiations Bulletin* 12: 702. http://enb.iisd.org/download/pdf/enb12702e.pdf

OECD (2015). *Multilateral Aid 2015: Better Partnerships for a Post-2015 World*, Paris: OECD.

Patz, R. and Goetz, K. H. (2017). Changing Budgeting Administration in International Organizations: Budgetary Pressures, Complex Principals and Administrative Leadership. In M. W. Bauer, C. Knill, and S. Eckhard (eds.), *International Bureaucracy: Challenges and Lessons for Public Administration Research*. Basingstoke: Palgrave Macmillan, 123–150.

Seitz, K. and Martens, J. (2017). Philanthrolateralism: Private Funding and Corporate Influence in the United Nations, *Global Policy* 8 (supplement 5): 46–50.

Sollberger, K. (2020). Extraordinary Meeting of the Conference of the Parties to the Convention on Biodiversity and the Meetings of the Parties to the Cartagena and Nagoya Protocols: 16–19 and 25–27 November 2020, *Earth Negotiations Bulletin* 9: 753.

United Nations (2016). *Budgetary and Financial Situation of the Organizations of the United Nations System: Note by the Secretary-General, A/71/583\**. https://documents-dds-ny.un.org/doc/UNDOC/GEN/N16/350/40/PDF/N1635040.pdf?OpenElement

United Nations General Assembly (1946–1958). *Reports of the Committee on Contributions*. New York: UN General Assembly.

UNFCCC (2014). *Momentum for Change: 2013 Annual Report*. Bonn, Germany: UNFCCC. http://unfccc.int/files/secretariat/momentum_for_change/application/pdf/mfcannualreport2013-web.pdf

Wagner, L. M. and Mwangi, W. (2010). Be Careful What You Compromise For: Postagreement Negotiations within the UN Desertification Convention, *International Negotiation* 15 (3): 439–458.

# 7

# More Resources – More Influence of International Bureaucracies?

## *The Case of the UNFCCC Secretariat's Clean Development Mechanism Regulation*

KATHARINA MICHAELOWA AND AXEL MICHAELOWA

## 7.1 Introduction

Without an active secretariat, decisions under international treaties would often be ill-prepared, and an informed negotiation process would be much more difficult to achieve. Formally, secretariats are supposed to be neutral technocrats and not meant to influence democratic decision-making processes. In reality, however, things are usually different. In fact, it is almost impossible to provide "impartial information," since even the volume of the information provided and the way it is prepared and introduced into the debate generally have some political impact. This relates to what Barnett and Finnemore (1999: 708) have identified as the "irony of depoliticized appearance." An active secretariat has to act behind the scenes, "indeed in the corridors and hotel bars of conference venues" (Bauer 2006: 34), and this hidden and informal action may be a key determinant of any progress to be achieved.

At the same time, the secretariat's influence necessarily constrains the role of elected decision-making bodies. Influence thereby relates to both the design of policy outputs and the control of decision-making processes. Thus, secretariats need to strike a "delicate balance between the activism that is needed to make a difference and the risk of being perceived as questioning or even challenging specific interests of individual parties to the treaty," that is, objectionable political interference (Andresen and Skjærseth 1999: 7; Bauer 2006: 34). From a normative perspective, the role the secretariats should assume in this context depends on a number of context variables. These include the complexity of the problem that calls for the knowledge of specialized experts and the diversity of political preferences that call for a clear predominance of the democratic decision-making bodies and a less active role of the secretariat (see, e.g., Alesina and Tabellini 2007, 2008; Hawkins et al. 2006b). A number of studies exist that compare the influence of different secretariats along these lines. Biermann and Siebenhüner (2009) have provided a comprehensive discussion of different international environmental agreements.

This academic literature reflects a recent trend in the international relations literature to shift the focus to international bureaucracies as relevant independent actors, and not just acting on behalf of their member states (Barnett and Finnemore 1999, 2004; Hawkins et al. 2006a; Johnson 2013a, b; Johnson and Urpelainen 2014; Michaelowa, Reinsberg, and Schneider 2018), and a simultaneous trend within the economic theory of bureaucracy to consider a more realistic objective function for civil servants that significantly departs from the simple resource maximization perspective introduced by Niskanen (1971) (see, e.g., Alesina and Tabellini 2007: 173; Dewatripont, Jewitt, and Tirole 1999a, b).

However, most of these scholars seem to concentrate on the process of initial delegation: How much autonomy should states delegate to an international bureaucracy for a specific task, what safeguards should they impose to limit agency slack (both shirking and slippage), and how does the empirical difference observed between the responsibilities delegated to different international organizations reflect the theoretical (functionalistic) expectations about the extent of delegation?

In contrast, our analysis focuses on the behavior of international bureaucracies once they are established. Such studies are typically carried out from a sociological or anthropological perspective, which has now also found its way into political science (see Barnett and Finnemore 2004 for a general discussion; for an in-depth study of individual organizations see, e.g., Weaver 2008 for the World Bank). In contrast, Hawkins and Jacoby (2006) combine the detailed observation of international bureaucracies' activities with the principal–agent approach generally used within the economic theory of bureaucracy. By doing so, they adopt the assumption of international civil servants rationally following their own, independent, objectives. Hawkins and Jacoby illustrate their theoretical approach with the example of how the two main institutions of the European Convention of Human Rights (ECHR) changed the way of accepting cases and their court decisions in response to enlarged country membership and thus went clearly beyond their original mandate. The timing of events allows the authors to argue convincingly that it was the purposefully designed strategy of the international bureaucracy, rather than the state-led initial institutional design, that resulted in the considerable autonomy of these institutions.

Yet further empirical illustrations of the suitability of the principal–agent approach for the explanation of the behavior of international bureaucracies are rare. In the context of a case study on the International Monetary Fund, Gould (2006: 306 ff.) argues that the principal–agent approach is useful to predict the behavior of the principals but much less so to predict the behavior of the secretariat. She concludes that exclusively relying on principal–agent theory may well explain why an international bureaucracy is endowed with a certain level of autonomy but not in which way it will actually make use of this autonomy. Regarding environmental bureaucracies, Manulak (2017) uses a principal–agent framework to study the entire process from initial delegation to subsequent attempts by states to informally control the secretariat of the United Nations Environment Programme. Other work on environmental bureaucracies sometimes reflects upon dynamics within bureaucracies but does not create a systematic and comprehensive link to principal–agent theory.

In this chapter, we attempt to overcome these problems by combining the rational choice approach of the principal–agent framework with some of the more constructivist ideas of principal–agent interactions that may eventually lead to revisions of the principals' initial objectives. This goes beyond conventional principal–agent theory, which assumes a static set of (mutually conflicting) preferences for the agent and the principal. This new framework allows us to capture the interesting process of bureaucracies reinterpreting and redefining their rules – a process highlighted both in Barnett and Finnemore (2004) and in Hawkins and Jacoby (2006).

We believe that over and above a more realistic and detailed definition of the international bureaucracies' objectives (as compared to pure budget maximization), it is these dynamics in the interaction between the agent and the principal that should allow us to derive more precise predictions about concrete activities of international civil servants within the general rational choice approach of the principal–agent model. Moreover, rather than studying bureaucratic behavior in general, we focus on the analysis of bureaucratic strategies triggered by resource growth – a rather typical situation for many international organizations.

Empirically, we illustrate our arguments with the example of the secretariat of the United Nations Framework Convention on Climate Change (UNFCCC) (henceforth simply called "the Secretariat") and notably its relatively technical branch responsible for international market mechanisms, especially the Clean Development Mechanism (CDM). In such an area that is politically much less contested than, for example, emission reduction commitments, we expect the greatest chance to observe the development of the independent role of an international bureaucracy. In addition, this particular case study enables us to combine both econometric analysis based on quantitative data on resource growth, the range of delegated activities, and actual policy decisions (see Michaelowa and Michaelowa 2017) and more in-depth qualitative analysis based on document analysis and interviews, on which we focus here. Our data are unique in that they allow us to measure a resource increase exogenous to deliberate decisions by the principal. This is crucial for our empirical identification strategy.

In addition, the initial role of the Secretariat has been relatively well researched, providing us with a sound basis for our analysis. In particular, Depledge (2005, 2007) and Yamin and Depledge (2004) provide a detailed account and discussion of the Secretariat's tasks and activities. Moreover, Busch (2009) analyzes the UNFCCC Secretariat within the comparative theoretical framework for different environmental treaty secretariats provided by Biermann and Siebenhüner (2009).

In this chapter, which complements and expands upon the more quantitative analysis undertaken in Michaelowa and Michaelowa (2017), we analyze whether the Secretariat has become more powerful over time, whether and how it started to directly influence policy processes, and how these developments are linked to the growth in financial resources. To do so, we first provide a more detailed and general theoretical framework in Section 7.2. Next, in Section 7.3 we provide an introduction to the specific UNFCCC case study and notably a description of the unexpected resource flow to a certain area of the Secretariat due to the CDM. In Section 7.4 we analyze two cases where the Secretariat increased its influence on rule-setting. In Section 7.5 we discuss how the drying up of the CDM market has led to an even stronger tendency for "top down" rule-setting as well as "Parkinson's law"-style responses by the Secretariat, while in Section 7.6 we provide our conclusion.

## 7.2 Principal–Agent Interactions When International Secretariats Grow: A Theoretical Framework

As pointed out by Alesina and Tabellini (2007: 173), there is no established standard model of bureaucratic behavior so far. In this context, a central problem appears to be the appropriate specification of the arguments in the bureaucratic objective function. Despite considerable discussion leading to a variety of different models in the rational-choice literature, consensus appears to emerge on at least some issues.

First, it is generally agreed that the sole consideration of the general budget (Niskanen 1971) or the discretionary budget (resources minus cost) (Migué and Bélanger 1974; Niskanen 1975), that is, the consideration of resources as the only argument bureaucracies really care about, does not provide us with a sufficiently realistic theoretical framework to predict the specific development of diverse international bureaucracies. Barnett and Finnemore (1999: 706) mention the opposition of North Atlantic Treaty Organization staff against political expansion plans in the late 1990s as one example that calls for other, supplementary, arguments.

Second, while budgetary concerns should not be the only ones considered, this does not mean that they are unimportant. Despite the obvious oversimplification, the focus on resource expansion objectives within the rational-choice framework has been able to provide useful explanations to a variety of phenomena observed within international organizations (see, e.g., Vaubel 1991, 2006). Even Barnett and Finnemore (1999: 706) acknowledge that "there is good reason to assume that organizations care about their resource base and turf." They insist, however, that resources should not be the only motivation taken into account as bureaucrats will usually weight them against other objectives. If other objectives are seriously taken into account, their striving for resources may also lose some of its negative connotation. In fact, at least to some extent, resources may then simply represent a means for the bureaucracy to reach other, socially more highly valued, ends.

Third, the existing literature has already brought up good candidates to complement the list of variables in the bureaucratic objective function. According to many authors (e.g., Alesina and Tabellini 2007, 2008; Barnett and Finnemore 1999, 2004) bureaucrats want to show technical excellence in their field of expertise. This may be related to the goal of acceptance and prestige within their professional community (Alesina and Tabellini 2007), to related career concerns (Dewatripont, Jewitt, and Tirole 1999a, b; Holmstrom 1982), or simply to their normative commitment to the services (i.e., eventually, the public good) delivered by their organization (see, e.g., Hawkins and Jacoby 2006: 223). Indeed, since bureaucrats are not hired at random, but from a community of people who self-selected into this specific field of activity in the first place, they should also be expected to be more dedicated to this field than the average citizen (Häfliger and Hug 2012). In fact, many of the functionalist explanations for the very existence of international organizations focus on the states' willingness to delegate tasks to an organization with particularly strong substance-related preferences, so that it would help them to overcome problems of credible commitment, notably under potentially time-inconsistent preferences (Hawkins et al. 2006b: 18–19).

Another typical objective bureaucracies are expected to value is autonomy or power (Barnett and Finnemore 2004; Hawkins et al. 2006b; Vaubel 1991). We follow Hawkins et al. (2006b: 8) in defining autonomy more broadly than discretion, a term that is used to refer to autonomy within the restricted area of explicitly delegated activities. Autonomy implies some freedom for independent action and thereby, eventually, some influence on actual policy outcomes.

To some extent, it is a logical implication of the principal–agent model itself that bureaucrats should value autonomy. The reason is that autonomy allows them to follow their own objectives as opposed to those of the principals. From this perspective, however, autonomy is purely instrumental, while it has been frequently considered as an objective in its own right (see, e.g., Barnett and Finnemore 2004; Hawkins and Jacoby 2006). Indeed while one might conceive autonomy as instrumental to, for example, the objective to show the bureaucracy's technical competence or, alternatively, to the objective to enlarge the bureaucracy's resource base, one might also conceive these two other objectives as instrumental to autonomy. As Barnett and Finnemore (2004: 21) put it:

Bureaucracy is powerful and commands deference, not in its own right, but because of the values it claims to embody and the people it claims to serve. IOs [International organizations] cannot simply say, "we are bureaucracies, do what we say." To be authoritative, ergo powerful, they must be seen to serve some valued and legitimate social purpose, and, further, they must be seen to serve that purpose in an impartial and technocratic way using their impersonal rules.

All in all, we have thus identified three widely agreed objectives that may all be final goals but also means to achieve some of the other goals. Bureaucratic utility ($U^B$) could thus be specified as a function of these three terms: $U^B = f(autonomy, excellence, resources)$, whereby the exact functional form, that is, the weighting of the different arguments and the way they interact, remains to be specified depending on the more specific context within which the analysis takes place. For instance, in a politically contested and nationally salient context, striving for autonomy should be less relevant because further delegation may a priori be unrealistic. The opposite should be true in a context in which policy decisions depend primarily on technical assessments. Moreover, the dynamics between the different objectives can be expected to depend on context too. If, as in the empirical example discussed later, we observe an exogenous (and substantial) increase in financial resources, the expected dynamics would be that these resources will be used to promote the other two objectives (and then, only in a second step, further resource growth).

While the list of objectives considered in the bureaucratic utility function may omit some variables that could reasonably be added (again depending on context), it should generally provide a sound basis for a reasonable prediction of bureaucratic behavior within international organizations. The above specification also shows that the rational choice framework with an explicitly defined utility function does not preclude altruistic behavior, or bureaucratic behavior that effectively serves the internationally agreed goals of the organization (be it based on altruistic motivations or not). Thus, while we need to make some assumptions about the bureaucratic objective function, we do not necessarily need to make normative judgments in order to predict the developments of international bureaucracies within the framework of a principal–agent model.

By setting up a more specific bureaucratic objective function we respond to one problem identified in the literature with respect to the application of the principal–agent approach to the identification of bureaucratic behavior (Barnett and Finnemore 1999; Gould 2006). However, this leads to yet another problem: Integrating autonomy in the objective function (and all the related dynamics suggested earlier) is at odds with the static formulation of the traditional principal–agent framework. In such a static framework, striving for autonomy cannot lead anywhere because the rules are defined once and for all. As Hawkins and Jacoby (2006) convincingly argue, de facto, many international bureaucracies do succeed in obtaining greater autonomy over time. They do so both by reinterpreting existing rules (and gradually changing accepted practices) and by convincing the principals that rules might have to be changed (i.e., by convincing them to formally delegate more autonomy). Barnett and Finnemore (2004) also make a strong point that the rules themselves may be endogenous to bureaucratic activity.

While principal–agent models have been adjusted to include multiple principals and several hierarchical levels of principals and agents (e.g., citizens delegating a task to their respective governments who in turn delegate it to an international bureaucracy), they typically ignore that the principals' interests and priorities may change over time (Stone 2011: 26). The adjustment of the principals' beliefs about how much authority they should delegate would typically rather be discussed in a constructivist framework. Yet the two can be fruitfully married here, since the adjustment of beliefs may well be a very rational choice by principals, notably in the context of imperfect information that the principal–agent model supposes anyway, and thus in line with the general assumptions of this model.

We believe that it is important to highlight the break with the static version of the principal agent–model because these dynamics are essential to explain bureaucratic behavior. While Hawkins and Jacoby (2006) do not explicitly mark this theoretical break, the actual importance of these dynamics in their work is omnipresent, notably in their empirical analysis.

Delegation of autonomy is not a once and for all decision but is subject to constant adjustments either through the reinterpretation or through the formal revision of existing rules. Principals decide (and redecide) based on an ongoing optimization between the reduction of their own workload and improved outcomes due to the use of bureaucratic expertise on the one hand and the cost of engaging the bureaucracy on the other hand. The consideration of cost includes not only the direct financial cost of maintaining an international civil service but also the political cost of potentially undesirable bureaucratic decisions. Principals are induced to accept or even actively promote greater bureaucratic autonomy (i) if they receive relevant information (related, for instance, to a large international crisis or a change in public awareness), (ii) if new external resources (e.g., from the private sector) become available to cover some of the cost, or (iii) if change is obfuscated, for example, if decisions about relevant procedural rules are hidden in the midst of complicated technicalities, or simply, if change is gradually creeping in.

This third channel is driven by the bureaucracy itself, notably by showing its excellent technical capabilities or by generating trust by giving itself a very technocratic and apolitical appearance. Both external factors (i) and (ii) can complement and facilitate the bureaucracy's strategy in this respect. For instance, the bureaucracy can make use of new information by interpreting it in a way that makes its expertise more desirable, or it can use new resources to enhance its capacity and the actual and/or perceived quality of its services. For instance, by hiring competent staff, the principal may be more easily convinced to leave relevant tasks in the responsibility of the secretariat.

In addition, if staff are sufficiently large and experienced, it may be more easily able to convince the principal of the predominance of technical and process-related issues even regarding decisions that are, de facto, less technical than political. Finally, it may take the time to carefully draft propositions in a way to increase its own procedural rights without anyone noticing, or it may simply overwhelm the principals with so many issues to decide upon that they cannot help but delegate some of these decisions (or their preparation) back to the secretariat. Generally speaking, with increasing resources, a bureaucracy disposes of greater means to increase pressure for more autonomy and greater means to show competence. This should enable the bureaucracy to extend the compelling offer to reduce the workload of principals, thereby enhancing its own freedom of action and its impact on actual decision-making.

Apart from the dynamics we introduce into the model, there is one more way in which we wish to deviate from the traditional principal–agent framework. The principal–agent model usually adopts the normative perspective of the principals. In our context, this does not necessarily make sense because the principal–agent relationship we observe is only a subset of a wider hierarchical principal–agent framework, and the interests of national delegates at international organizations may themselves largely deviate from the interests of the ultimate principal, namely the population in the member countries. Thus, we could observe situations in which the national delegates willingly delegate more authority to the secretariat to reduce their own workload, while the general public would have preferred these issues to be decided at a political level. In other situations, political positions of national delegates may be driven by narrow vested interests, in which case the general public would prefer an international bureaucracy dedicated to the delivery of the global public good (i.e., to the central goal of the organization) to take over responsibility.

In brief, this implies that unless we take the whole picture into account, a normative judgment cannot be made. In the following, we thus concentrate on a positive analysis and only hint to the potential normative implications here and there without the intention to be conclusive in this respect.

## 7.3 The UNFCCC as an Empirical Case Study

The UNFCCC was agreed upon by the governments participating at the United Nations Conference on Environment and Development (Rio Conference) in 1992 and entered into force in February 1994. So there is now thirty years of experience with the work of the Secretariat. Our study covers the time period including 2016, that is, until the entry into force of the Paris Agreement. The dynamics after 2016 have changed considerably compared to the preceding period and would warrant a separate assessment. For example, from 2021 onward the CDM, which has provided a significant share of the Secretariat's funding (see discussion later), has been replaced by new international carbon markets under Article 6 of the Paris Agreement (see Ahonen et al. 2022).

## *Initial Delegation to the UNFCCC Secretariat and Prospects for Further Development*

The existing literature on the Secretariat refers to its initial set-up and the first few years of its existence. According to Depledge (2005: 70 ff.) the Secretariat's activities were purposefully constrained by the member states to ensure a minimum of technical functionality while avoiding any kind of political interference such as experienced with other environmental treaties. The Secretariat's activities include the provision of relevant logistics, procedural management, advice to the relevant presiding officers, technical advice in general, drafting text, and the facilitation of informal discussions. Depledge (2005: 73) underscores that on the basis of this mandate, the Secretariat indeed chose a strictly apolitical "behind the scenes approach" as opposed to the approaches of other treaty secretariats such as the early ozone regime. Similarly, according to Busch (2009: 251),

> the climate secretariat is a 'technocratic bureaucracy' that has not had any autonomous political influence.... It has not promoted its own agenda or pursued specific approaches but has responded to requests of parties. It has functioned as an important and valuable but passive information hub in the climate regime that does not autonomously interfere with any political, scientific, or public discourses.

In his study, which refers to the early to mid-1990s, he concludes that the UNFCCC Secretariat is one of the least powerful among nine environmental treaty secretariats under comparison (see also Bauer, Busch, and Siebenhüner 2009).

Biermann and Siebenhüner (2009) provide a broader study, in which these nine environmental treaty secretariats are compared. The analysis is based on a comprehensive theoretical framework distinguishing between the cognitive, normative, and executive influence of these secretariats (Biermann et al. 2009). Within this theoretical framework, Busch (2009) identifies problem structure as the central argument for the strong constraints imposed on the UNFCCC Secretariat. Climate change is a politically complex issue on which scientific results continue to evolve and which cannot be solved by technical or administrative means. The salience of the problem as well as the cost of public action is high, and national interests are widely diverging.

When political positions diverge and when the issues are nationally very salient, and compliance to adverse decisions very costly, member countries will not give decisions out of hand and instead keep them directly within political decision-making arenas (Biermann et al. 2009). Stone (2011: 23) and more recently Manulak (2017) provide a complementary rationale for this behavior. They argue that member countries know that autonomy delegated to international bureaucracies can always be used by powerful countries to exert informal influence via these bureaucracies. Thus, granting autonomy to bureaucracies may effectively mean granting more power to some members relative to others. When issues are politically highly contentious, member countries may thus try to avoid such delegation in the first place. According to Stone (2011), a similar argument applies when there is little international consensus about the fundamental purposes of the organization.

In contrast, issues of high technical complexity call for stronger delegation to an international bureaucracy, because such tasks require considerable time and expertise, and autonomous decisions of political committees as well as close monitoring of the bureaucracy become very expensive.

While overall the political element clearly dominates the technical element in international climate policy, it should be noted that the general field covered by the climate negotiations hides a lot of specific, and indeed sometimes also quite technical, issues. Correspondingly, the Secretariat is no monolithic block, and within the Secretariat, some areas dealing with more technical issues may more easily gain autonomy than others – which may lead to quite imbalanced developments within the bureaucratic structure.

However, if at all we expect any change over time, this would require that bureaucrats have somehow been able to convince the principals that more delegation is advantageous for them. According to our theoretical framework, the bureaucracy may try to gradually enhance its autonomy by exerting some influence on the principals. Until recently, the literature suggested, however, that in the case of the UNFCCC, the "straitjacket" imposed on the Secretariat, which rules out any proactive or independent role, also influences its culture in a way to make such developments rather improbable (Bauer, Busch, and Siebenhüner 2009: 178; Busch 2009: 261). Depledge (2005: 73) even concluded that any influence the Secretariat may have critically depends on its invisibility. None of these authors seemed to believe that the Secretariat's role would see significant changes in the future and that the Secretariat would itself even try to make it change. However, more recently van Asselt and Zelli (2018), Hickmann et al. (Chapter 3), and Well et al. (Chapter 4) assert that the Secretariat has become more proactive.

On the basis of our theoretical framework we empirically assess how the inflow of resources and information contributed to the Secretariat's more proactive role.

A relevant flow of information that would challenge the balance of interest leading to the initial delegation decision cannot be observed. While scientific outcomes have led to a stronger international consensus on the reality of anthropogenic climate change, views on the implications widely diverge, and there is no consensus whatsoever on the responsibilities and commitments to be taken over by individual members (Gupta 2012). The strong political differences between countries are further illustrated by the failure of the Copenhagen conference in 2009 and the postponement of any decision about further steps to 2015. From this perspective, it can thus not be expected that the Secretariat should have become more powerful.

However, we do observe a significant externally induced growth in resources that was, in addition, fully unexpected by member countries and the Secretariat alike. This will allow us, in the following sections, to specifically analyze the impact of resource growth on the dynamics of our model.

### *The Development of UNFCCC Resources*

Let us start by considering some descriptive statistics regarding the overall development of UNFCCC resources, along with some initial interpretations. Following the UNFCCC's entry into force in February 1994, it took two years to establish the Secretariat. The first budget, available for the biennium 1996/1997, shows expenditures of about USD 4.5 million per year. Until 2015, it increased twentyfold, with particularly strong absolute increases in 2007–2010 (see Figure 7.1). The general growth in resources is mirrored by the growth in staff, which increased from 34 (20 professional and 14 administrative positions in 1995; Depledge 2005: 63) to 558.5 (346 professional-level and 212.5 administrative posts) in 2015. In 2016, both budget and staff numbers fell significantly.

The initial major growth phase seems to be related to the preparation of the Marrakech Accords (November 2001) that provided the detailed specifications for

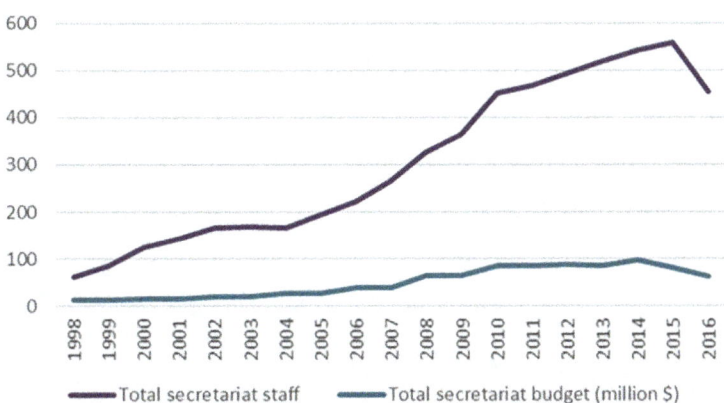

Figure 7.1 Development of financial and human resources of the UNFCCC Secretariat
*Note:* The Secretariat budget includes the core budget and all trust funds except the Trust Fund for Participation in the UNFCCC process, which includes only transitory positions. When figures are reported other than for a full calendar year, they are annualized assuming an even spread of expenditures over the year.
*Source:* See Table 7.A1

the implementation of the Kyoto Protocol agreed upon in 1997. The second major growth phase could then be related to the Kyoto Protocol's actual entry into force in February 2005, which implied, in particular, the regular assessment of the parties' greenhouse gas emissions and the evaluation of methodologies and projects submitted in the context of market mechanisms (trade in emission reduction certificates). Finally, the specific rise in 2009 could be related to the expected tasks in the context of the Copenhagen conference in late 2009, which was supposed to bring about an agreement on the follow-up to the Kyoto Protocol after the end of its first commitment period. These interpretations are only partially plausible, however, when looking at the more specific distribution of funds within the Secretariat.

In the period studied, the most important trust fund was the fund for the CDM. The CDM was a market mechanism under the Kyoto Protocol that allowed generation of emission credits (Certified Emission Reductions, CERs) from mitigation projects in developing countries. CERs could be used by industrialized countries to fulfil their Kyoto emission targets. CDM projects had to be formally registered by the CDM Executive Board (EB), which is supported by Secretariat staff in its decision-making. For this service, a fee was charged by the Secretariat both for registration of CDM projects and for CER issuance. The inflow of money through this source was much higher than originally predicted (see Michaelowa and Michaelowa 2017: 251). This led to the accumulation of a surplus, which developed over time as shown in Figure 7.2, as reallocation of funds to other areas of Secretariat activities was impossible (Michaelowa and Michaelowa 2017: 252).

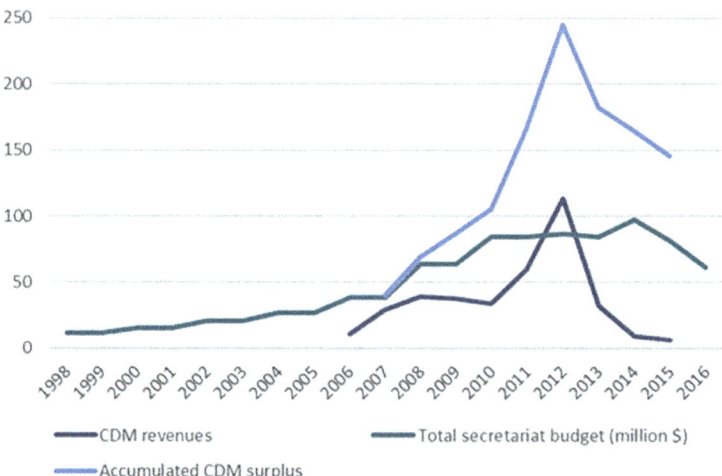

Figure 7.2 CDM revenues for the secretariat and the development of the accumulated surplus over time (million USD)
*Source:* See Table 7.A1

Much of the steep rise in total Secretariat staff and expenditure after 2007 can be explained by the resources for the CDM. Note in particular that the strong increase in total expenditure between 2007 and 2010 observed in Figure 7.1 and not explained so far coincides exactly with the increase of CDM revenues. Thus, the Copenhagen meeting in November 2010 may be one explanation for the latter, but the market-driven rise of the CDM along with the related resource growth appears to be the predominant one. Obviously, the two explanations are not mutually exclusive as there may be a happy cooccurrence of needs and available means. However, as the political focus of this major international conference on climate change was on renewed emission reduction commitments, rather than CDM-related activities, there is some doubt about how CDM-related financial resources could have been used in this context.

Owing to the failure of the Copenhagen conference, the willingness of key countries to engage in acquisition of emission credits fell significantly in the early 2010s. A key example is the European Union, which prohibited the import of certain kinds of CERs and also introduced a maximum quota, which was attained around 2013. The CER price fell from EUR 12 in early 2011 to EUR 0.4 in early 2013 and stayed below EUR 1 until the end of the period. After a peak in CDM registration in late 2012, which was due to an EU deadline for credit eligibility, registration slowed to a trickle (see Michaelowa, Shishlov, and Brescia 2019).

As resources were restricted to being used in the context of the CDM, the revenue increase until 2010 directly translated into staff increases in the Secretariat's CDM department (see Figure 7.3). Afterward, staffing was kept constant while the revenue surplus increased further until 2012 (see Figure 7.2). This was due to problems in hiring sufficiently knowledgeable staff. During that period the Secretariat outsourced

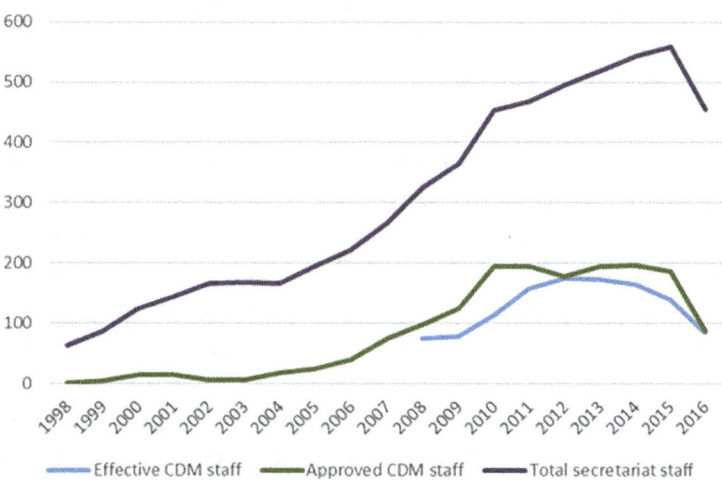

Figure 7.3 CDM versus total secretariat staff
*Source:* See Table 7.A1

substantial work to external consultants. When the CDM market crisis hit (see discussion in Section 7.5 and Michaelowa, Shishlov, and Brescia 2019), initially consultancy assignments were scaled back while the Secretariat's effective staff started to decrease slowly. Only in late 2015, approved staff levels were slashed massively.

Summing up the developments between the entry into force of the UNFCCC in 1994 and of the Paris Agreement in 2016, the most striking features are thus not only the strong increase in overall financial and human resources but also the increasing dominance of the CDM part within these resources, which is institutionally protected by the international agreement not to use income from the CDM fees for any other purposes. As shown in Figure 7.2, in 2012 the CDM revenues exceeded the entire budget of the Secretariat. Only after 2014, the role of the CDM in the Secretariat's work declined, long after the market had come to a standstill.

The theoretical framework derived in Section 7.2 suggests that bureaucrats would make use of an important external resource flow like the CDM revenues to influence the principals in order to obtain more autonomy. The following subsection will apply the general theoretical framework to the specific context of the UNFCCC to lay the groundwork for the case studies in Section 7.4.

### *The Secretariat as Rule-Setter in the Case of a Resource Glut*

Our theoretical framework summarized in Sections 7.1 and 7.2 suggests that external resource flows will indeed provide an opportunity for the international bureaucracy to influence its principals more effectively. In addition, the external resources allocated to the Secretariat make the international bureaucracy weigh less heavily on the member countries' budget, which directly influences the principals' balance between the cost and benefits of delegation. For both reasons, principals can be expected to rely more heavily on the bureaucracy, to delegate more responsibilities, and to accept the related increase in resources (if relevant). With increased autonomy and resources, which can be used to hire additional skilled professionals, the international bureaucracy can then even further influence the principals, so that we would expect a self-reinforcing effect of the initial external shock whose dynamics might fade out only after some time, when a new equilibrium is reached. Such changes in the influence of the Secretariat could also influence its own self-perception, its organizational culture, and its confidence in pushing for even further autonomy. This would be a specific example for the self-reinforcing dynamics of the initial resource flow. We will not be able to precisely disentangle the individual pathways driving these developments (e.g., whether it is the impact on organizational culture, on the quality of bureaucratic services, or on staff's self-confidence that is primarily responsible for these self-reinforcing dynamics). In the following section, we discuss two cases where the Secretariat expanded its influence on rule-setting. For a quantitative analysis of the early increase in CDM resources and the influence on rule-setting, especially in the context of the checking of quality of project documentation, see Michaelowa and Michaelowa (2017).

## 7.4 Changes in CDM Rules and Regulations Concerning the Secretariat's Freedom of Action

The stronger role of the Secretariat can be best exemplified in two specific areas: (i) decisions about baseline and monitoring methodologies for potential future CDM projects and (ii) the issue of a standardization of baseline methodologies (standardized baselines). In both cases the stronger role was made possible by the increased availability of expert manpower at the Secretariat, which enabled the Secretariat to argue that it would provide faster and more high-quality methodology-related work than what the "bottom-up" external expert review process could provide. A third case on rules for request for review of problematic project proposals was discussed in depth in Michaelowa and Michaelowa (2017: 255–256).

**Case 1: Baseline and Monitoring Methodologies**

Baseline and monitoring methodologies were key to determine the amount of emission credits of a CDM project. They thus directly influenced the amount of money a country received for the export of emission credits and were thus commercially important. Project developers could submit methodology proposals,[1] which were evaluated by the Methodologies Panel (Meth Panel) and then submitted to the CDM EB, which normally followed the Meth Panel's recommendation. Traditionally, methodology submissions were evaluated by independent desk reviewers chosen by the Meth Panel. According to information from the EB, the increasing role of the Secretariat is due to the EB's assessment that the Meth Panel could not handle the increasing number of methodologies. In June 2007, a preassessment of proposals by the Secretariat was introduced (see decision EB 32, Annex 13). While one Meth Panel member selected by the Secretariat would check this (para 7), the Secretariat would develop a draft recommendation (para 14). From February 2010, the Secretariat could skip the independent desk review (see decision EB 52, Annex 9, para 18) if supported by two members of the Meth Panel chosen by the Secretariat itself. It is likely that the Secretariat did not choose overly critical Meth Panel members if it wanted to push a methodology. Moreover, from 2010 onward, the Secretariat started to engage in methodology development on its own initiative, an area previously reserved to external developers. From late 2012 the Secretariat

---

[1] When speaking about "methodologies" in this chapter we only refer to so-called "large-scale methodologies," that is, methodologies for projects above a certain size threshold (at 15 MW for renewable energy, 15 GWh of annual savings for energy efficiency projects, and 60 000 t $CO_2$ annual reductions for all other project types).

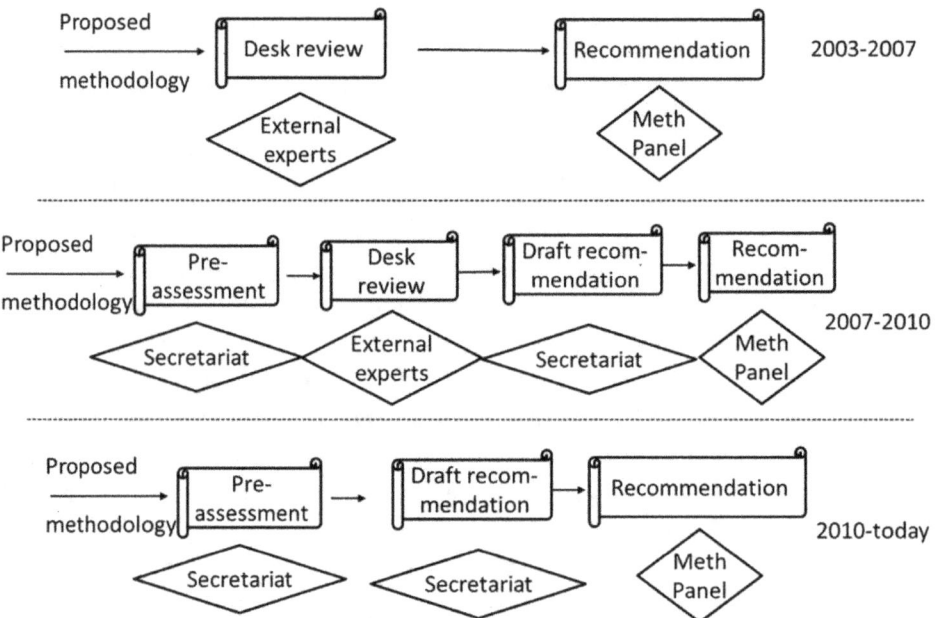

Figure 7.4 Changes in procedures for approval of baseline and monitoring methodologies over time
*Note:* The final decision on a proposed methodology is always taken by the EB

could officially propose methods, while at the same time being the only institution systematically involved in the quality assessment of the methodologies (see Figure 7.4, and summary of the rulemaking changes in UNFCCC 2013a). As no private CDM project developer was willing to invest into methodology development after the CDM price crash, all methodologies submitted since 2012 were developed by the Secretariat or its Regional Collaboration Centres (see Section 7.5). The shift from independent review to secretariat-led rulemaking likely reduced the overall stringency and conservativeness of methodologies.

As experience with methodology application accumulated, flaws became visible and project developers were able to ask for a revision of approved methodologies. Usually, they aimed to reduce the transaction costs linked to the use of a methodology as well as its stringency. Traditionally, the Meth Panel prepared the recommendation whether to engage on a revision of a methodology while the Secretariat did the completeness check of the revision request. This was similar to the traditional division of labor between the Secretariat and the Registration and Issuance Team (RIT) for CDM projects. In October 2007, EB 35 (Annex 13, para 8) introduced drafting of the recommendation by the Secretariat. From November 2010 onward (decision EB 54, Annex 2), the Secretariat was able to initiate methodology revisions on its own initiative (para 7). It could then hire outside consultants for preparation of the draft recommendation but needed the approval of the Meth Panel chair (para 14), before selecting one or two Meth Panel members for review. Subsequently, the Secretariat could call for public comments and change the methodology draft if it found the comments relevant. Here again, the approval of the Meth Panel chair was required (para 21d). The Secretariat could also trigger "editorial amendments," which just needed to be approved by the Meth Panel chair and entered into force automatically unless an EB member objected (paras 29–30). This meant that the Secretariat was able to control the whole revision process if supported by the Meth Panel chair. As noted by one of our interviewees, this would have made sense to speed up processes if the Meth Panel as a whole had difficulties to find a consensus. Yet, it clearly implied an increase in responsibilities for the Secretariat. Here, the shift from the project developer-led revisions to secretariat-led revisions likely increased the overall stringency and conservativeness of methodologies.

**Case 2: Standardized Baselines**

A strong influence of the Secretariat was also visible in the process of standardizing baselines (i.e., scenarios against which emission reductions by CDM projects had to be assessed). Initially, the EB had asked the Secretariat to develop proposals in consultation with the Meth Panel. However, the rules agreed in September 2011 (decision EB 63, Annex 28) effectively allowed the Secretariat to bypass the Meth Panel with regard to evaluating submissions of standardized baselines (EB 68, Annex 32, paras 15, 16, and 22) and to send the baseline directly to the EB (para 27). While there was still an obligation to include two Meth Panel members assessing the Secretariat's proposal (para 23), this safeguard could be easily weakened by selecting members disposed favorably to the Secretariat's proposal.

In November 2011, the EB approved guidelines for standardized baselines and related performance benchmarks developed by the Secretariat (EB 65, Annex 23). A dispute between the Meth Panel and the Secretariat arose regarding the appropriateness of these guidelines. After a long debate, this dispute resulted in an official "information note" sent by the Meth Panel to the EB in November 2012 (UNFCCC 2012d). Thereby, for the first time, an official committee of the UNFCCC questioned the quality of the Secretariat's work.

In a similar way, a dispute broke out in the context of the determination of the benchmarks used within the standardized baselines. The contested numerical values were hidden in a document innocuously named "Work programme on standardized baselines" (EB 65, Annex 22, para 10). Again, the Meth Panel openly criticized the Secretariat's approach as documented in another official "information note" for the EB (UNFCCC 2012c).

Since 2013, the standardization drive of methodologies by the Secretariat accelerated. Within three years, standardized shares of nonrenewable biomass – relevant

Table 7.1 *Standardized baselines developed with Secretariat support*

| Year | Power | Waste | Cookstoves | Agriculture | Energy efficiency | Forestry |
|---|---|---|---|---|---|---|
| 2013 | 2 | | | | 1 | |
| 2014 | 2 | | | | | |
| 2015 | 3 | 5 | 2 | 1 | | |
| 2016 | 5 | 5 | 1 | 2 | | 1 |
| 2017 | 1 | | 4 | | | |
| 2018 | 2 | 1 | | | | |

*Source:* UNFCCC (2019)

for efficient cookstove projects – were calculated for thirty-four countries. That this was not done according to market demand is shown by the fact that these parameters were used only by forty-one CDM activities – thus on average by just one activity (UNFCCC 2016a). Moreover, forty standardized baselines were developed on a country level for various sectors with the support of Regional Collaboration Centres until the end of 2018, as shown in Table 7.1.

One of the national delegates to the UNFCCC articulated concern about the strong concentration of resources in the CDM-related part of the Secretariat. During the interview, he underscored that, in general, the Secretariat's inputs and advice have been extremely useful: "There have been a few cases where the Secretariat put things on the agenda, which created a lengthy process. But in other cases, if the Secretariat had been followed, a lot of time could have been saved. Overall, the negative cases are infinitesimal as compared to the positive side." Despite this highly positive overall appreciation of the Secretariat's work, he asserted that the accumulation of resources in the CDM part of the Secretariat clearly required restructuring.

This perspective on resources was challenged by other respondents. They believed that it makes sense for the Secretariat to concentrate resources on an area that is more technical and less politically contested. It was mentioned that a certain financial buffer for the CDM was actually intended to overcome "bad times." At the same time, another respondent pointed at the experience from Joint Implementation (JI), where the number of projects and thus income from fees had not risen in the same way as for the CDM (see Table 7.A1 for JI staffing, revenue, and budget). In this area, the Secretariat had been much less active in promoting new rules and processes and in proposing increases in its own responsibilities to the corresponding political committee.

Overall, the two cases show how over time and in line with the increase in staff resources the Secretariat took over significant responsibilities in the development and revision of baseline and monitoring methodologies. As such methodologies were critical in determining the monetary revenues of activities under the CDM, the Secretariat was now able to influence which types of activities would be able to generate emissions credits under which circumstances.

### 7.5 Secretariat Reactions to the CDM Market Crisis

Even after the CER market price had crashed, the Secretariat still kept a large staff force active in the CDM department. Given the reduced inflow of projects, initially Secretariat-led development of rules was intensified as discussed in Section 7.4. A number of baseline and monitoring methodologies were developed by Secretariat staff, and a large number of methodologies were revised. Moreover, outreach to various stakeholders was undertaken to increase the attractiveness of the CDM, for example, setting up a web platform for voluntary cancellation of CERs, which was launched in September 2015 (UNFCCC 2016c). Before the emergence of national-level policy instruments such as the Korean emissions trading scheme and the Colombian carbon tax that accepted CER cancellation certificates in 2017, the platform was used only to a small extent. After the emergence of international climate finance institutions such as the Green Climate Fund, the Secretariat had attempted, in vain, to market CERs to these institutions.

Moreover, the regulatory documents defining the project cycles and other key regulatory steps were aggregated in overall documents, a work that may have pleased legal practitioners but did not have any immediate impact on the use of the mechanism.

Last but not least, a loan scheme to support project developers was launched in April 2012. By late June 2016, 191 applications had been received, with 78 loan agreements signed totaling USD 6.21 million (UNFCCC 2016c: 15). The scheme was closed at the end of 2018. UNFCCC (2018a) states that while 63 loans with a total volume of USD 3.7 million had actually been paid out, only about half were expected to be fully repaid while more than 20 loans were likely to be written off completely, with the remaining loans likely to be partially repaid. Already in 2012 it would have been clear to any observer that this money could never be paid back in the difficult market situation with CER prices close to zero.

The most visible activity was the setting up of Regional Collaboration Centres in all major world regions. This started immediately after the market crisis: The center in Lomé, Togo, started in January 2013, followed by Kampala, Uganda (May 2013), St. George's, Grenada (July 2013), and Bogota, Colombia (August 2013). Another one in Bangkok followed in September 2015. While initially the focus of the centers was on developing CDM project pipelines, they did become general capacity-building entities, supporting knowledge transfer in the context of the Paris Agreement and its new concepts such as nationally determined contributions (NDCs), as shown directly in the statement "Set up to spread the benefits of the clean development mechanism (CDM), the RCCs have broadened their role since adoption of the Paris Agreement on climate change in December 2015, supporting the development and implementation of countries' nationally determined contributions to climate action under that agreement" (UNFCCC 2017a: 1). As per this new mandate, the key activities of Regional Collaboration Centres in 2018 were to support the development of measurement, reporting, and verification systems and the elaboration of studies on domestic carbon pricing policy instruments and NDC partnership plans (UNFCCC 2018b).

## 7.6 Conclusions

The UNFCCC Secretariat was able to mobilize an unexpected volume of revenues from the CDM, a market mechanism under the Kyoto Protocol that proved surprisingly attractive to the private sector between 2005 and 2011. The inflow of over USD 350 million within a decade led to a rapid expansion of staffing at the Secretariat and a tendency to take over rule-setting under the mechanism. We provide evidence through case studies on rules for development of baseline and monitoring methodologies as well as standardization of such methodologies. This complements evidence found through regression analysis by Michaelowa and Michaelowa (2017). Our assessments confirm theoretical considerations that an international bureaucracy tends to take over tasks from its member governments as soon as its resources increase.

The collapse of the CDM market from 2012 onward initially led to a "hibernation" attitude of the Secretariat, which only slowly laid off staff and even increased Secretariat-led rule-making, despite lack of activity on the market. Standardization of methodologies was undertaken and the rulebook streamlined significantly. However, this can be partially seen as a manifestation of Parkinson's Law, as a number of activities were undertaken that clearly did not have significant benefits, such as loans to project developers and brokerage activities to find buyers for emission credits. Only after five years of crisis was a serious downscaling of staff undertaken. At the same time, activities of remaining staff were tacitly reoriented to support negotiations under the Paris Agreement and its implementation, for example in the context of Regional Collaboration Centres that could no longer support identification of projects.

## Acknowledgments

We thank Liliana Andonova, Carsten Hefeker, and Tana Johnson for their very thoughtful comments that have helped us to substantially improve this chapter. Moreover, we thank all those that shared their precious time with us during our interviews, and without whom this chapter would have been much less rich in concrete examples.

# Annex

Table 7.A1 *Overview of the UNFCCC Secretariat's development as well as of its CDM- and JI-related activities*

| Phase | Year | Executive Secretary | Total Secretariat expenditure (million USD) | CDM Trust Fund expenditure (million USD) | Accumulated CDM surplus (million USD) | JI budget (million USD) | Total staff | CDM staff | JI staff | CDM methodo-logies[a] | CDM projects[b] | JI projects[c] | Key events |
|---|---|---|---|---|---|---|---|---|---|---|---|---|---|
| 0 | 1992 | Zammit Cutajar | NA | | | | NA | | | | | | Rio Conference agrees on UNFCCC |
| 0 | 1993 | Zammit Cutajar | NA | | | | NA | | | | | | |
| 1 | 1994 | Zammit Cutajar | NA | | | | NA | | | | | | UNFCCC entry into force |
| 1 | 1995 | Zammit Cutajar | NA | | | | 34 | | | | | | Berlin mandate |
| 1 | 1996 | Zammit Cutajar | 4.5 | | | | NA | | | | | | |
| 1 | 1997 | Zammit Cutajar | 4.5 | | | | 47 | | | | | | Kyoto Protocol |
| 2 | 1998 | Zammit Cutajar | 10.6 | | | | 63 | 1.5 | | | | | |
| 2 | 1999 | Zammit Cutajar | 10.6 | | | | 86 | 5 | | | | | |
| 2 | 2000 | Zammit Cutajar | 15.6 | | | | 124 | 5 | | | | | |

Table 7.A1 (cont.)

| Phase | Year | Executive Secretary | Total Secretariat expenditure (million USD) | CDM Trust Fund expenditure (million USD) | Accumulated CDM surplus (million USD) | JI budget (million USD) | Total staff | CDM staff | JI staff | CDM methodologies[a] | CDM projects[b] | JI projects[c] | Key events |
|---|---|---|---|---|---|---|---|---|---|---|---|---|---|
| 2 | 2001 | Zammit Cutajar | 15.6 | | | | 144 | 5 | | | | | Marrakech Accords (November) |
| 3 | 2002 | Waller-Hunter | 20.8 | | | | 165.5 | 6 | | | | | |
| 3 | 2003 | Waller-Hunter | 20.8 | | | | 168.5 | 6 | | 36 | | | |
| 3 | 2004 | Waller-Hunter | 26.7 | | | | 165.5 | 17 | | 50 | 2 | | |
| 3 | 2005 | Waller-Hunter | 26.7 | | | | 194.5 | 25 | | 76 | 141 | | Kyoto Protocol entry into force (February) |
| 4 | 2006 | de Boer | 38.6 | | | 2.0 | 221.5 | 40 | 3 | 75 | 443 | 1 | |
| 4 | 2007 | de Boer | 38.6 | | | 3.1 | 265.5 | 75 | 9 | 52 | 567 | 28 | Secretariat summary note for registration requests and recommendations for methodology revisions |
| 4 | 2008 | de Boer | 63.8 | 18.3 | 39.9 | 3.1 | 325.5 | 97 | 10 | 66 | 717 | 13 | |

| | | | | | | | | | | | | | |
|---|---|---|---|---|---|---|---|---|---|---|---|---|---|
| 4 | 2009 | de Boer | 63.8 | 18.3 | 69.3 | 3.5 | 364.5 | 125 | 14 | 61 | 727 | 11 | Copenhagen conference fails (December) |
| 5 | 2010 | Figueres | 84.1 | 31.7 | 86.4 | 3.4 | 452.5 | 194 | 11 | 23 | 756 | 11 | Review procedure changed |
| 5 | 2011 | Figueres | 84.1 | 31.7 | 104.6 | 2.2 | 467.5 | 194 | 12 | 33 | 961 | 10 | Rules for standardized baselines introduced |
| 5 | 2012 | Figueres | 86.9 | 44.7 | 166.2 | 1.9 | 494.5 | 195 | 11 | 18 | 3268 | 2 | Top-down methodologies introduced |
| 5 | 2013 | Figueres | 84.2 | 35.0 | 245.1 | 1.7 | 519 | 196 | 11 | 14 | 222 | | |
| 5 | 2014 | Figueres | 97.5 | 40.9 | 182.3 | 1.2 | 543 | 195 | 6 | 4 | 154 | | |
| 5 | 2015 | Figueres | 81.5 | 29.3 | 164.9 | 1.0 | 558.5 | 187 | 4 | 3 | 86 | | Paris Agreement (December) |
| 6 | 2016 | Espinosa | 61.5 | 14.5 | 145.1 | 0.8 | 455 | 87 | 4 | 4 | 57 | | |

[a] Annual numbers according to open comments date for large-scale methodologies and submission date for forestry and small-scale methodologies.
[b] Submission for registration.
[c] Submission for determination (track 2). Only those that are overseen by the Secretariat.

*Source*: UNFCCC (1996, 1997, 1999a, b, c, 2000, 2001, 2002, 2003a, b, 2004, 2005, 2006, 2007, 2008, 2009, 2010, 2011, 2012a, 2013a, b, 2014a, b, 2015a, b, c, 2016a, b, c, 2017b), UNEP DTU (2017)

## References

Ahonen, H.-M., Kessler, J., Michaelowa, A., Espelage, A., and Hoch, S. (2022). Governance of Fragmented Compliance and Voluntary Carbon Markets under the Paris Agreement, *Politics and Governance* 10 (1): 235–245. DOI: 10.17645/pag.v10.

Alesina, A. and Tabellini, G. (2007). Bureaucrats or Politicians? Part I: A Single Policy Task, *American Economic Review* 97: 169–179.

Alesina, A. and Tabellini, G. (2008). Bureaucrats or Politicians? Part II: Multiple Policy Tasks, *Journal of Public Economics* 92: 426–447.

Andresen, S. and Skjærseth, J. B. (1999). *Can International Environmental Secretariats Promote Effective Co-operation?* Paper Presented at the United Nations University's International Conference on Synergies and Co-ordination between Multilateral Environmental Agreements, July 14–16, Tokyo. http://archive.unu.edu/inter-linkages/1999/docs/Andresen.PDF

Barnett, M. and Finnemore, M. (1999). The Politics, Power, and Pathologies of International Organizations, *International Organization* 53 (4): 699–732.

Barnett, M. and Finnemore, M. (2004). *Rules for the World: International Organizations in Global Politics*, Ithaca, NY: Cornell University Press.

Bauer, S. (2006). Does Bureaucracy Really Matter? The Authority of Intergovernmental Treaty Secretariats in Global Environmental Politics, *Global Environmental Politics* 6 (1): 23–49.

Bauer, S., Busch, P.-O., and Siebenhüner, B. (2009). Treaty Secretariats in Global Environmental Governance. In F. Biermann, B. Siebenhüner, and A. Schreyögg (eds.), *International Organizations in Global Environmental Governance*, London: Routledge, 174–192.

Biermann, F. and Siebenhüner, B. (eds.) (2009). *Managers of Global Change: The Influence of International Environmental Bureaucracies*, Cambridge, MA: MIT Press.

Biermann, F., Siebenhüner, B., Bauer, S., et al. (2009). Studying the Influence of International Bureaucracies. A Conceptual Framework. In F. Biermann and B. Siebenhüner (eds.), *Managers of Global Change: The Influence of International Environmental Bureaucracies*, Cambridge, MA: MIT Press, 37–74.

Busch, P.-O. (2009). The Climate Secretariat: Making a Living in a Straitjacket. In F. Biermann and B. Siebenhüner (eds.), *Managers of Global Change: The Influence of International Environmental Bureaucracies*, Cambridge, MA: MIT Press, 245–264.

Depledge, J. (2005). *The Organization of International Negotiations: Constructing the Climate Change Regime*, London: Earthscan.

Depledge, J. (2007). A Special Relationship: Chairpersons and the Secretariat in the Climate Change Negotiations, *Global Environmental Politics* 7 (1): 45–68.

Dewatripont, M., Jewitt, I., and Tirole, J. (1999a). The Economics of Career Concerns, Part I: Comparing Information Structures, *Review of Economic Studies* 66 (1): 183–198.

Dewatripont, M., Jewitt, I., and Tirole, J. (1999b). The Economics of Career Concerns, Part II: Application to Missions and Accountability of Government Agencies, *Review of Economic Studies* 66 (1): 199–217.

Gould, E. (2006). Delegating IMF Conditionality: Understanding Variations in Control and Conformity. In D. Hawkins, D. Lake, D. Nielson and M. Tierney (eds.), *Delegation and Agency in International Organizations*, Cambridge: Cambridge University Press, 281–311.

Gupta, J. (2012). Negotiating Challenges and Climate Change, *Climate Policy* 12: 630–644.

Häfliger, U. and Hug, S. (2012). *International Organizations, Their Employees and Volunteers and Their Values*, Paper presented at the IPSA XXII World Congress of Political Science, Madrid, July 8–12.

Hawkins, D. and Jacoby, W. (2006). How Agents Matter. In D. Hawkins, D. Lake, D. Nielson and M. Tierney (eds.), *Delegation and Agency in International Organizations*, Cambridge: Cambridge University Press, 199–228.

Hawkins, D., Lake, D., Nielson, D., and Tierney, M. (eds.) (2006a). *Delegation and Agency in International Organizations*, Cambridge: Cambridge University Press.

Hawkins, D., Lake, D., Nielson, D., and Tierney, M. (2006b). Delegation under Anarchy: States, International Organizations, and Principal-Agent Theory. In D. Hawkins, D. Lake, D. Nielson and M. Tierney (eds.), *Delegation and Agency in International Organizations*, Cambridge: Cambridge University Press, 3–38.

Holmstrom, B. (1982). *Managerial Incentive Problems: A Dynamic Perspective. Essays in Economics and Management in Honor of Lars Wahbleck*, Vol. 66, Helsinki: Swedish School of Economics. Reprinted in *Review of Economic Studies* (1999): 169–182.

Johnson, T. (2013a). Institutional Design and Bureaucrats' Impact on Political Control, *Journal of Politics* 75 (1): 183–197.

Johnson, T. (2013b). Looking beyond States: Openings for International Bureaucrats to Enter the Institutional Design Process, *Review of International Organizations* 8 (4): 499–519.

Johnson, T. and Urpelainen, J. (2014). International Bureaucrats and the Formation of Intergovernmental Organizations: Institutional Design Discretion Sweetens the Pot, *International Organization* 68 (1): 177–209.

Manulak, M. (2017). Leading by Design: Informal Influence and International Secretariats, *The Review of International Organizations* 12 (4): 497–522.

Michaelowa, A., Shishlov, I., and Brescia, D. (2019). Evolution of International Carbon Markets: Lessons for the Paris Agreement, *WIREs Climate Change* 10: e613, DOI: 10.1002/wcc.613.

Michaelowa, K. and Michaelowa, A. (2017). The Growing Influence of the UNFCCC Secretariat on the Clean Development Mechanism, *International Environmental Agreements: Politics, Law and Economics* 17: 247–269.

Michaelowa, K., Reinsberg, B., and Schneider, C. (2018). The Politics of Double Delegation in the European Union, *International Studies Quarterly* 62: 821–833.

Migué, J.-L. and Bélanger, G. (1974). Towards a General Theory of Managerial Discretion, *Public Choice* 17: 24–43.

Niskanen, W. (1971). *Bureaucracy and Representative Government*, Chicago: Aldine-Atherton.

Niskanen, W. (1975). Bureaucrats and Politicians, *Journal of Law and Economics* 8: 617–643.

Parkinson, N. (1957). *Parkinson's Law, and Other Studies in Administration*, Cambridge, MA: Riverside Press.

Stone, R. (2011). *Controlling Institutions: International Organizations and the Global Economy*, Cambridge: Cambridge University Press.

UNEP DTU (2017). *CDM Pipeline*. www.cdmpipeline.org

UNFCCC (1996). *Financial Performance of UNFCCC: Contributions and Expenditures in 1996 and Forecast for the Biennium 1996–1997*, FCCC/CP/1996/7, Bonn.

UNFCCC (1997). *Administrative and Financial Matters*, FCCC/SBI/1997/10, Bonn.

UNFCCC (1999a). *Administrative and Financial Matters*, FCCC/SBI/1999/3, Bonn.

UNFCCC (1999b). *Income and Budget Performance in the Biennium 1998–1999*, FCCC/SBI/1999/10, Bonn.

UNFCCC (1999c). *Programme Budget for the Biennium 2000–2001*, FCCC/SBI/1999/4, Bonn.

UNFCCC (2000). *Income and Budget Performance in the Biennium 2000–2001*, Interim Report as at 30 June 2000, FCCC/SBI/2000/8, Bonn.

UNFCCC (2001). *Income and Budget Performance in the Biennium 2000–2001*, FCCC/SBI/2001/16, Bonn.

UNFCCC (2002). *Income and Budget Performance as at 30 June 2002*, FCCC/SBI/2002/11, Bonn.
UNFCCC (2003a). *Interim Financial Performance for the Biennium 2002–2003. Income and Budget Performance as at 30 June 2003*, FCCC/SBI/2003/12, Bonn.
UNFCCC (2003b). *Programme Budget for the Biennium 2004–2005*, FCCC/SBI/2003/5/Add.1, Bonn.
UNFCCC (2004). *Income and Budget Performance as at 30 June 2004*, FCCC/SBI/2004/13, Bonn.
UNFCCC (2005). *Budget Performance for the Biennium 2004–2005 as at 30 June 2005*, FCCC/SBI/2005/13, Bonn.
UNFCCC (2006). *Budget Performance for the Biennium 2006–2007 as at 30 June 2006*, FCCC/SBI/2006/15, Bonn.
UNFCCC (2007). *Budget Performance for the Biennium 2006–2007 as at 30 June 2007*, FCCC/SBI/2007/19, Bonn.
UNFCCC (2008). *Budget Performance for the Biennium 2008–2009 as at 30 June 2008*, FCCC/SBI/2008/10, Bonn.
UNFCCC (2009). *Budget Performance for the Biennium 2008–2009 as at 30 June 2009*, FCCC/SBI/2009/11, Bonn.
UNFCCC (2010). *Budget Performance for the Biennium 2010–2011 as at 30 June 2010*, FCCC/SBI/2010/13, Bonn.
UNFCCC (2011). *Budget Performance for the Biennium 2010–2011 as at 30 June 2011*, FCCC/SBI/2011/16, Bonn.
UNFCCC (2012a). *Budget Performance for the Biennium 2012–2013 as at 30 June 2012*, FCCC/SBI/2012/23, Bonn.
UNFCCC (2012b). *CDM Management Plan 2012*, EB 66 Report, Annex 2, Bonn.
UNFCCC (2012c). *Information Note on Proposed Draft Guidelines for Determination of Baseline and Additionality Thresholds for Standardized Baselines Using the Performance-Penetration Approach*, CDM-MP58-A20, Bonn.
UNFCCC (2012d). *Information Note on Way Forward with Guidelines for the Establishment of Sector Specific Standardized Baselines*, CDM-MP58-A18, Bonn.
UNFCCC (2013a). *Annual Report of the Executive Board of the Clean Development Mechanism to the Conference of the Parties Serving as the Meeting of the Parties to the Kyoto Protocol*, FCCC/KP/CMP/2013/5, Bonn.
UNFCCC (2013b). *Budget Performance for the Biennium 2012–2013 as at 30 June 2013*, FCCC/SBI/2013/14, Bonn.
UNFCCC (2014a). *Annual Report of the Executive Board of the Clean Development Mechanism to the Conference of the Parties Serving as the Meeting of the Parties to the Kyoto Protocol*, FCCC/KP/CMP/2014/5, Bonn.
UNFCCC (2014b). *Budget Performance for the Biennium 2014–2015 as at 30 June 2014*, FCCC/SBI/2014/10, Bonn.
UNFCCC (2015a). *Annual Report of the Executive Board of the Clean Development Mechanism to the Conference of the Parties Serving as the Meeting of the Parties to the Kyoto Protocol*, FCCC/KP/CMP/2016/4, Bonn.
UNFCCC (2015b). *Budget Performance for the Biennium 2014–2015 as at 30 June 2015*, FCCC/SBI/2015/13, Bonn.
UNFCCC (2015c). *Joint Implementation Two-Year Business Plan and Management Plan 2016–2017*, JI-JISC37-A01, Bonn.
UNFCCC (2016a). *Annual Report of the Executive Board of the Clean Development Mechanism to the Conference of the Parties Serving as the Meeting of the Parties to the Kyoto Protocol*, FCCC/KP/CMP/2015/5, Bonn.

UNFCCC (2016b). *Budget Performance for the Biennium 2016–2017 as at 30 June 2016*, FCCC/SBI/2016/13, Bonn.

UNFCCC (2016c). *CDM Two-Year Business Plan 2016–2017 and Management Plan 2016*, CDM-EB87-A01-INFO, Bonn.

UNFCCC (2017a). *UNFCCC Regional Collaboration Centres Catalyse Climate Action*. https://unfccc.int/files/secretariat/regional_collaboration_centres/application/pdf/rcc_brochure_eng_final_2017.pdf

UNFCCC (2017b). *CDM Management Plan 2017*, CDM-EB92-A01-INFO, Bonn.

UNFCCC (2018a). *Annual Report of the CDM Loan Scheme*, CDM-EB100-AA-A08, Bonn.

UNFCCC (2018b). *Regional Collaboration Centres 2018 Highlights*. https://unfccc.int/sites/default/files/resource/RCC%20Highlights%202018.pdf

UNFCCC (2019). *Approved Standardized Baselines*. https://cdm.unfccc.int/methodologies/standard_base/2015/sb4.html

van Asselt, H. and Zelli, F. (2018). International Governance: Polycentric Governing by and beyond the UNFCCC. In A. Jordan, D. Huitema, H. van Asselt, and J. Forster (eds.), *Governing Climate Change: Polycentricity in Action?*, Cambridge: Cambridge University Press, 29–46.

Vaubel, R. (1991). The Political Economy of the International Monetary Fund: A Public Choice Analysis. In R. Vaubel and T. Willet (eds.), *The Political Economy of International Organizations: A Public Choice Approach*, Boulder, CO: Westview Press, 204–244.

Vaubel, R. (2006). Principal-Agent Problems in International Organizations, *Review of International Organizations* 1 (2): 125–138.

Weaver, C. (2008). *Hypocrisy Trap: The World Bank and the Poverty of Reform*, Princeton: Princeton University Press.

Yamin, F. and Depledge, J. (2004). *The International Climate Change Regime: A Guide to Rules, Institutions and Procedures*, Cambridge: Cambridge University Press.

# 8

# The Marrakech Partnership for Global Climate Action

*Democratic Legitimacy, Orchestration, and the Role of International Secretariats*

KARIN BÄCKSTRAND AND JONATHAN W. KUYPER

## 8.1 Introduction

Over the past three decades, the secretariats of international organizations (IOs) have faced a dilemma. On the one hand, these bodies are tasked with helping to solve some of the most pressing issues facing the international community, including the spread of infectious diseases, spiraling conflict in civil war zones, a lack of equitable trade and investment between countries, and the adverse impacts of climatic change. On the other, these bodies often face a lack of fiscal resources, finite staff capacity, and the inability to issue hard or legally binding regulation. Due to this tension, international secretariats – and the bureaucracy that they comprise – have increasingly turned toward *orchestration* as a mode of governance. Orchestration is defined as "the mobilization of an intermediary by an orchestrator on a voluntary basis in pursuit of joint governance goals" (Abbott et al. 2016: 719). As described by Abbott et al. (2015), it is an attenuated type of governance: Governors, acting as orchestrators (O), seek to direct the governed as targets (T) through third-party intermediaries (I) (see also Chapter 3). This indirect form of governance differs from a traditional principal–agent (P–A) relationship in which governors set firm mandates and boundaries for the governed, seeking to reward compliance and sanction deviation.

At a Global Climate Action High Level Event at the United Nations Framework Convention on Climate Change (UNFCCC) twenty-sixth Conference of Parties (COP26) in Glasgow in November 2021, the UN Secretary-General António Guterres stressed that a decarbonized and resilient world meeting the 1.5°C goal requires an "all hands on the deck" approach involving governments, business, and civil society. He emphasized the participation gap between the Global North and South with regard to nonstate climate commitments and announced the decision to establish a high-level expert group to develop standards to measure and evaluate net zero commitments by nonstate actors (UN News 2021). The online Global Climate

Action Portal (GCAP), which consists of voluntary climate actions from almost 33,000 actors, was relaunched at COP26 to track the progress of climate commitments. This represents a shift toward emphasizing the democratic legitimacy of transnational climate governance action: to promote broadened participation and representation of vulnerable stakeholders as well as strengthened transparency and accountability mechanisms of the "groundswell" of nonstate climate action. These shifts, and the relationship between the secretariat, orchestration, strategies, and democratic legitimacy, are the topics of this chapter.

As noted in Chapter 1, many scholars today acknowledge that orchestration has become a prevalent activity in international relations as IOs – seeking to tackle transnational problems without the ability to exercise hard law – turn toward soft forms of governance and steering. Recent studies of the UNFCCC Secretariat demonstrate that it, by itself and in tandem with other actors, especially after the adoption of the Paris Agreement, engages in orchestration (see Chapter 3; Hickmann and Elsässer 2020). To date, the lion's share of work on orchestration has been either conceptual (expounding the constituent features of the concept), exploratory (showing the mechanisms through which it works), or explanatory (focusing on the effectiveness and/or problem-solving ability of the activity).

In this chapter, we analyze the normative dimensions associated with orchestration, such as democratic values related to participation, accountability, transparency, and deliberation. We note that orchestration, for all its importance, triggers a set of legitimacy questions. That is, the indirect mode of governance muddles who should be held accountable for which actions, to which set of standards, and which agents have the right to demand said accountability. We treat this as a democratic issue. Our core argument is that the practice of orchestration engenders a democratic duty. Orchestrators – be it intergovernmental organizations or states – need to ensure that their own actions, and those of intermediaries, are democratically legitimated by affected stakeholders, including both targets and additional actors implicated in the orchestration relationship.

In making this argument, the chapter is divided into four sections. First, we advance the orchestration concept and introduce a novel theoretical element focusing on meta-intermediaries. Second, we turn to democratic theory and argue that orchestration, by virtue of the usage of public authority, triggers a democratic demand. Building on earlier work (Bäckstrand and Kuyper 2017) we develop this argument and contend that a "democratic values" approach represents a useful way to evaluate the accountability and legitimacy of orchestrators. Next, we apply this argument to orchestration by the UNFCCC Secretariat, notably through the Marrakech Partnership for Global Climate Action (GCA), which is a multistakeholder framework for accelerating climate action among nonstate actors in line with the Paris Agreement's goals of decarbonization by 2050. While previous research on orchestration of the UNFCCC has predominantly focused on the effectiveness nonstate action in the GCA (Hale et al. 2021; Hsu et al. 2015), we show how and why nonstate climate action requires democratic legitimation. To that end, we apply the democratic values approach and demonstrate that substantial democratic deficits exist. The final section concludes by discussing the intrinsic and instrumental importance of evaluating orchestration through a democratic legitimacy lens and the implications for international secretariats.

## 8.2 Orchestration and Global Governance

International organizations have emerged as key players in the governance of different issue areas in world politics. These IOs are established to solve global collective action problems that sovereign states in isolation cannot manage due to complexity, a lack of information, or free-riding that might undermine problem-solving efforts. In the early post-World War II era, IOs were largely the handmaidens of states, particularly powerful ones. States gave mandatory financial contributions to IOs and, in return, received both formal and informal power concerning the direction and operation of that organization. International secretariats then derived their formal mandate from states through a classic principal–agent relationship.

However, in recent years, this model has eroded in a more polycentric world of complex and hybrid multilateralism. Today, the resources and funding allocated to international secretariats are constrained by states in highly selective ways (Graham 2017; see also Chapter 6). Moreover, it has become clear that international secretariats often have an independent set of preferences that may or may not coincide with those of the states that empower them. Secretariats are often populated by bureaucrats that want to solve collective action problems in line with their own normative vision, are granted mandates by states to tackle issues on their own, or seek to influence state preferences directly through interactions with delegates (Chapters 3 and 4). Literature has also shown that secretariats tackle problems in ways they see fit, either by stretching mandates through agency slack (see also Chapter 5) or by carving out new space to act entrepreneurially by sourcing public and private finance directly.

In many of these instances, secretariats are then turning toward orchestration as a way to engage other actors – intermediaries – in shared goals and projects. Although states (Hale and Roger 2014) such as Sweden (Nasiritousi and Grimm 2022), regions (Chan et al. 2019), and cities (Gordon and Johnson 2017) engage in orchestration, international secretariats from the European Union, the World Trade Organization, the G20, the World Health Organization, and the International Labour Organization have also adopted this strategy (Abbott et al. 2015). Orchestration allows secretariats to use the public authority granted to them by states to engage in problem-solving.

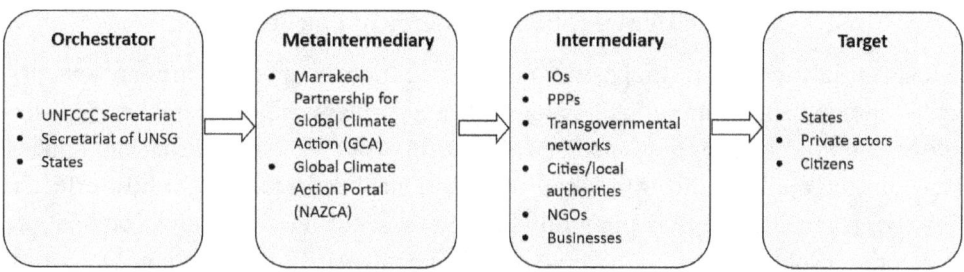

Figure 8.1 The O–I–T relationship

As noted earlier, and illustrated in Figure 8.1, orchestration can be conceptualized in terms of an Orchestrator–Intermediary–Target (O–I–T) relationship. Here, *orchestrators* – such as IO secretariats – seek to mobilize *intermediaries* – such as nonstate actors, other IOs, or actors in transgovernmental networks – on a voluntary basis to impact *targets* in pursuit of a governance goal. Orchestration is indirect and soft as the IO addresses the ultimate targets – such as consumers, states, firms, or the general public – via intermediaries and because the orchestrator lacks hard control over this chain. Through this mode of governance, the orchestrator grants material and ideational resources to the intermediary party who can attempt to pursue its goals without binding restrictions from the orchestrator. Building upon Abbott et al. (2015), we also note that orchestrators often employ "meta-intermediaries" – institutional mechanisms (such as the GCAP) that group together and organize intermediaries – to engage in orchestration.

Given this description, we argue that the UNFCCC Secretariat fits well with the orchestration model (see also Chapter 3 for the same line of argument). In recent years, the UNFCCC has displayed a relatively limited set of governance capacities in terms of staff and budget compared to other IOs. However, it has increasingly engaged in the orchestration of nonstate actors at formal events, such as at COPs and the Intersessionals in Bonn. Second, the UNFCCC has been willing to work with a wide range of intermediaries in this pursuit: transgovernmental networks, civil society, scientists, and investors. Finally, as we will discuss later, efforts by the UNFCCC to tackle climate change through setting up meta-intermediaries – the GCAP,[1] the Lima–Paris Action Agenda (LPAA), which was subsequently transformed to the GCA. These are institutional mechanisms orchestrated by the UNFCCC Secretariat – frequently in tandem with other IOs – to mobilize and catalyze the efforts of intermediaries to enhance mitigation efforts, scale up adaptation actions, and harness finance.

---

[1] Previously called the Non-state Actor Zone for Climate Action.

## 8.3 The Democratic Legitimacy of Orchestration

Most initial work on orchestration focused on theory-building and theory-development: specifying the core concepts and showing its analytical and explanatory potential in different cases. More recent work has shown that the concept does indeed apply to different issue areas and has begun looking at how effective this governance strategy is in mobilizing actors, securing compliance, and solving collective problems (see, e.g., Hale and Roger 2014). In this chapter, we seek to ask a different question: Is orchestration democratically legitimate? In order to make this question relevant, we have to substantiate three issues. First, what is democratic legitimacy? Second, why should it apply to issues of orchestration? And, third, who holds a duty to be democratically legitimate and, inversely, who has a right to exercise democratic control over said duty holder?

On the first question, we place the content of this chapter in a broader context concerning the normative structure of international politics. In recent years, a general recognition has emerged in both academia and policy practice that the power and authority exercised by actors beyond/across national boundaries is normatively problematic. Within the confines of the nation-state, the basic institutional structure of domestic society assigns the relevant rights and duties to different agents, distributes the appropriate burdens and benefits across society, and defines legitimate and illegitimate forms of power. Beyond the state, a lack of such basic institutional mechanisms means that questions of justice and legitimacy arise. Increasingly scholars debate who owes what to whom, in what order or magnitude, and which decision-making procedures should determine those allocations.

Although orchestration undoubtedly also triggers questions of distributive justice and individual/group rights, we focus on whether the decision to engage in orchestration – the procedure and institutional rules it entails – is legitimate (for earlier and similar arguments about public–private partnerships, see Bäckstrand 2008). In broad terms, we assume that decision-making should be appropriately constrained and rendered democratically accountable to the relevant set of agents. This is cashed out in different terms, depending on the type of legitimacy one is concerned with. For instance, those interested in *sociological* legitimacy care about whether decision-making procedures, and their outcomes, are viewed as acceptable by some audience. Those interested in *normative* legitimacy, as we are, care about whether decision-making procedures, and their outcomes, live up to some *ex ante* desirable virtue. In this category, different forms of democracy are employed to form a baseline of normative legitimacy. For instance, liberal democrats care about whether power promotes or undermines individual autonomy, neorepublican democrats are concerned with the exercise – and constraint – of arbitrary power, and deliberative democrats focus on the justificatory quality of decision-making (Habermas 1996).

We choose not to take a side in these contending debates. Rather, we suggest that these different virtues can matter in a broad democratic theory of legitimate public power. As such, and following most recent work, we adopt a *democratic values* approach to judging the legitimacy of decision-making (Dingwerth 2014; Kuyper 2014). Following several different models of democracy, we suggest that the legitimacy of decision-making can be determined by looking at whether the exercise of public power (i.e., decision-making) is *participatory, accountable, and transparent*, as well as *deliberative*. What do we mean by these values? And how can they be operationalized?

*Participation* means that those impacted by the exercise of authority should have the opportunity and ability to be involved in how that authority is wielded. This entails, following other liberal theories, an equal capacity to shape the rules, laws, and regulations that will impact their lives. We note that equality of participation may often necessitate forms of representation as individuals cannot always be directly involved in decision-making processes. National representatives or "nonelected representatives" (interest groups, nongovernmental organizations [NGOs], etc.) can all help connect individuals with sites of authority (Macdonald 2008). Precisely how equal participation is secured will and should vary depending upon the institutional scheme in need of democratic regulation, in this case orchestration.

*Accountability*, in a democratic sense, means that those impacted by decision-making should have the right to hold power wielders "to a set of standards, to judge whether they have fulfilled their responsibilities in light of these standards, and to impose sanctions if they determine that these responsibilities have not been met" (Grant and Keohane 2005: 29). This criterion, following neo-republican conception of democracy, gives implicated individuals the opportunity to hold decision-makers at different levels of governance accountable for their actions and stop the arbitrary exercise of authority that can undercut individual autonomy. Operative accountability mechanisms provide an ex ante incentive for decision-makers to take consideration of how impacted parties will react to decisions being made in their name.

For accountability to be meaningful, *transparency* is required. Transparency is here conceptualized as the disclosure of actions taken by public actors and institutions. Said disclosure should be offered to those bound up by decisions. Although it does not require third-party monitoring, transparency is often promoted and enhanced by demands for information. Several scholars have claimed that an overabundance of transparency can also limit a public's ability to discern and view relevant information (Peixoto 2013). We agree with this, and suggest that if publics find transparency procedures either obfuscatory or misleading, then accountability measures are necessary (this is why we tackle accountability and transparency together).

Finally, *deliberation* provides those impacted by decisions with a rationale for how rules are being formulated and applied in various contexts (Habermas 1996). This value derives from the field of deliberative democracy that stresses the importance of providing reciprocal and generalizable arguments for how authority is exercised and how it is connected to the public use of reasoning. Reciprocity means that justification is mutually acceptable to parties in a deliberation, whereas generalizability connotes a set of reasons that could be shared by different parties due to shared institutional or moral structures. Deliberation also means that representatives of those impacted have an opportunity to put their reasons forward, have said reasons considered by decision-makers, and justify policies in light of those reasons.

Having now stipulated *how* an actor could be democratic, we must now say *why* it applies to the issue of orchestration and *who* should be democratically legitimated by which set of actors in orchestration relationships (though we have already touched upon both issues in the preceding discussion).

On the first issue, we have several reasons to think that orchestration triggers a democratic demand. In essence, orchestration is an explicit or implicit attempt to change the behavior of others. Specifically, orchestrators seek to use resources to mobilize and catalyze intermediaries in order to affect the actions of targets. In most general terms, democratic legitimacy requires a holder – one that must live up to a set of democratic standards – and a *demos* – one that is capable of exercising democratic restraint over the holder. There are very deep and complex debates about what *kinds of actions* trigger democratic demands (see, e.g., Goodin 2007). In fundamental terms, most democratic theorists agree that only certain activities demand democratic legitimacy. We categorize these types of activity into three groups: affectedness, significantly affectedness, and subjectedness.

In terms of affectedness, an actor should be democratically legitimate when they *affect* the interests of others (Goodin 2007). Those affected become the relevant demos and are given participatory, accountability, and deliberative rights over how the holder wields that power. Some scholars find affectedness too broad – that is, actors might often be only weakly affected by some action and therefore do not deserve democratic standing. These scholars stress that only those *significantly affected*' should be able to democratically constrain power wielders (see Macdonald 2008). Finally, other scholars also find this narrower conceptualization still too broad. Instead, they argue that only *subjection* to legal or coercive power requires democratic legitimation (see Abizadeh 2012). Across these three variants, the scope of the demos narrows as only certain types of actions trigger democratic demands.

Following other recent work, we take affectedness as the baseline. When an agent exerts power that affects the interests of others, that group becomes the demos with a corresponding democratic right to shape the exercise of that power.

It is clear on this metric that orchestration triggers a democratic demand. This is because orchestration, by its nature, is an effort to steer, mobilize, and nudge the actions of others. In other words, orchestration is an explicit attempt to use public power and authority to affect the interests of others.

We could have gone with a more restrictive version here, either the "significantly affected" or the "subjectedness" criteria. We stay with affectedness as it is arguably the most prominent in the literature on democratic theory (Goodin 2007; see also Koenig-Archibugi 2017). Moreover, we suggest that – while the scope of the demos would be narrowed on these competing views – both would still apply in the case of orchestration. On the "significantly affected" view, orchestration certainly does have the quantitative and qualitative capacity to dramatically shape the lives of individuals. For instance, Gordon and Johnson (2017) show how orchestration by city networks entails quite stringent rankings schemes, which in turn impacts the scale of mitigation, adaptation, and financing projects adopted within the orchestrated jurisdictions. On the subjectedness criterion, orchestration by states, cities, and even IOs has legal effect. That is, states, cities, and IOs (using power delegated or captured from states) employ public authority in crafting orchestration policy. While this might result in "soft" forms of steering and *facilitative* rather than *directive* orchestration (Hickmann et al. 2021), it is the employment of authority that subjects others that triggers a democratic demand. As Hale and Roger (2014) show, states use different forms of authority (material, epistemic, moral, relational) in orchestration processes. But ultimately the ability to do this and the resources used in orchestration are ultimately derived from public authority which does bind citizens. Thus, even on this most narrow view, there are good reasons to think that orchestration requires democratic legitimation.

So far, we have shown how to measure democratic legitimacy and why it applies to orchestration efforts. We have used the affectedness view to make this claim, though we believe our argument is compatible with "stronger" versions of democratic theory. Finally, we have to show who owes democratic standing and to whom. As should be clear, we argue that the orchestrator requires democratic legitimation. That is, the orchestrator should be democratically responsive in terms of participation, accountability (and transparency), and deliberative justification to those they affect. The affected demos is primarily the targets on the ground but may also be other actors that are implicated in the actions of the orchestrator.

While the intermediaries are also affected by the orchestrator, we want to suggest that – because intermediaries often join with the orchestrator voluntarily and in the pursuit of shared goals – their democratic claim against the orchestrator is diminished. Much more important is how the orchestrator is rendered democratically accountable to those affected "on the ground." Similarly, the orchestrator has a duty to those affected to ensure that the efforts by intermediaries are also democratically legitimated. By this we mean that intermediaries should be open to participation, accountability, and justification as their actions affect targets. However, the primary duty remains with the orchestrator: If intermediaries violate the democratic rights of those affected, the orchestrator should remove the resources granted to those intermediaries.

As such, we move forward with this conceptualization. In the following case, we show how the UNFCCC Secretariat, states, and High-Level Champions (HLCs) jointly have engaged in orchestration through the establishment of meta-intermediaries such as the Marrakech Partnership; how intermediaries are mobilized through this meta-framework; and how these efforts have affected targets and other actors (who become the demos). The normative task of determining appropriate standards for those involved in different forms of governance activities is particularly vital in the case of orchestration because, as Abbott et al. (2016: 727) note, this process obfuscates clear lines of accountability. Specifically – and correctly – they argue that orchestration "cuts the chain of electoral accountability because the orchestrator lacks hard control over intermediaries. Ultimately, intermediaries exercise their authority in an (externally) uncontrolled and unaccountable way." This, however, does not alter the fact that the decision to engage in orchestration, and its ongoing impact, affects targets and others. This, in turn, triggers a democratic right for that demos to ensure that orchestrators are responsive for the decision and process of orchestration, including the activities of intermediaries. The discussion we outline here of both the democratic values approach and the specification of who owes democratic accountability to whom thus provides a much-needed normative backdrop to the process of orchestration.

## 8.4 The GCA: Orchestration in the UNFCCC and Beyond

To show how our normative conceptualization of the link between democratic theory and orchestration applies, we turn toward an illustrative case study of the orchestration efforts of the UNFCCC Secretariat in alliance with other actors. We are not seeking to make definite conclusions about the democratic legitimacy of orchestration by the UNFCCC Secretariat in tandem with the HLCs but rather show how our framework can be applied. This should inform future work on the democratic legitimacy of orchestration efforts in global climate change and other issue areas of world politics.

The Paris Agreement, reached at COP21 in 2015, entails a changing role for the UNFCCC and its secretariat (Falkner 2016). Most centrally, the agreement cements the UNFCCC as an orchestrator of transnational nonstate and substate climate action primarily directed to mitigation although orchestration of adaption actions has increased (Chan and Amling 2019). That is, the UNFCCC Secretariat is now crucially involved in the mobilization of voluntary commitments by nonstate actors – or "nonparty stakeholders" – to achieve the goals of the Paris Agreement.[2] This focus on nonstate actor contributions solidifies a long-standing trend of engagement by the UNFCCC Secretariat. Prior to COP21, the LPAA – a joint undertaking by Peruvian and French COP presidencies, the Office of the UN Secretary-General, and the UNFCCC Secretariat – was formed to demonstrate the major advancements by nonstate actors. The UNFCCC Secretariat is responsible for managing the Non-state Actor Zone for Climate Action (NAZCA) online portal (renamed the Global Climate Action Portal) to showcase the efforts of different nonstate and substate actors as they commit to emissions reductions and adaptation actions, or provide climate finance.[3] The rationale for NAZCA in the run-up to COP21 in Paris was to have a back-up option showcasing the universe of transnational climate action had the intergovernmental negotiations failed to produce a treaty (Hale 2016).

The Paris Agreement then leaves the UNFCCC Secretariat as an orchestrator of state and nonstate action in two different ways. First, the 2016 COP21 decision accompanying the Paris Agreement (UNFCCC 2016, Decision 1/CP.21) reiterates the importance of the secretariat as an implementing body of the UNFCCC, organizing the COPs and Intersessionals (including high-level events and side events), coordinating the submission of nationally determined contributions (NDCs), convening the NAZCA/GCAP portal with data partners,[4] and facilitating the technical examination process (TEP). Second, the decision was also to establish two HLCs, tasked with interfacing the convention and voluntary and collaborative climate action. These HDCs provide guidance to the secretariat, help organize the COPs and Technical Expert Meetings (TEMs), and collaborate with the executive secretary of the UNFCCC and the COP presidents to bolster and catalyze nonstate climate action to 2020. We focus here on the role of these HLCs and their interaction with the secretariat to accelerate nonstate and substate climate action.

These HLCs operate on a rolling basis, with terms lasting two years and a new appointment being made annually to ensure continuity. The first two champions in 2016 were Laurence Tubiana, France's Climate Change Ambassador, and Hakima El Haite, Minister Delegate to the Minister of Energy, Mines, Water and Environment of Morocco. In 2017 it was again Hakima El Haite, joined by Inia B. Seruiratu, Minister for Agriculture, Rural and Maritime Development and National Disaster Management in the Republic of Fiji. In 2018 M. Tomasz Chruszczow,

---

[2] Before the Paris Agreement, the UNFCCC Secretariat engaged in orchestration through the Momentum for Change initiative established in 2011. Befitting the focus of this book, however, we hone in on the actions of the secretariat and HLCs.
[3] http://climateaction.unfccc.int/about
[4] Including the Carbon Disclosure Project (CDP), Climate Bonds Initiative, UNEP, Global Covenant of Mayors and the Climate Group.

Special Envoy for Climate Change from the Ministry of Environment in Poland, replaced Hakima El Haite. In 2019 Gonzalo Muñoz from Chile took over from Inia Seruiratu. At COP26, which was postponed more than a year to November 2021 because of the Covid-19 pandemic, the champions were Nigel Topping, the ex-executive director of the Carbon Disclosure Project, and Gonzalo Muñoz. The latter was replaced by Mahmoud Mohieldin (previous executive director of the International Monetary Fund) as the incoming champion for COP27 in Egypt. While the Marrakech Partnership was originally set up to mobilize pre-2020 action, COP25 in Madrid 2019 decided to extend the partnership as well as the role of the HLCs to 2025. Furthermore, states decided to establish standards for tracking the progress of nonstate climate action (UNFCCC Decision 1/COP25), including those launched at the UN Secretary-General's Climate Action Summit in New York 2019.

Champions – in collaboration with the secretariat – mobilize and orchestrate nonstate actor commitments through a variety of modes, including high-level events at COPs, the Climate Action Pathways, regional climate weeks, the Yearbook of Global Climate Action, and the Race to Zero, Race to Resilience, and Glasgow Finance Alliance for Net Zero campaigns. These streams are part of a broad push for enhanced action from 2016 to 2025 under the umbrella of the Marrakesh Partnership. The Marrakesh Partnership builds foundationally upon the LPAA and NAZCA, thus deepening the relationship between the HLCs, the secretariat, and the UN Secretary-General (Hale et al. 2021). For instance, initiatives mobilized by the HLC are to be included in NAZCA, and the GCA officially replaces LPAA as the central way to ratchet up ambition and to provide input the 2023 global stocktake of the Paris Agreement (Hsu et al. 2023). Overall, these various mechanisms as part of the GCA enable HLCs and the secretariat to orchestrate nonstate climate action.

In the next section, we ask whether the orchestration efforts by the HLC and the secretariat, in the form of the GCA, are democratically legitimate. Because separating cleanly between the secretariat and the HLC is very difficult empirically, we refer to the democratic legitimacy of the GCA, as it encompasses collaborative actions by both the champions.

## 8.5 Orchestration and Democratic Legitimacy: The GCA in Practice

In this final substantive section, we empirically assess the democratic legitimacy of the joint orchestration efforts by the UNFCCC Secretariat and the HLC when they work in collaboration. Overall, we find that the secretariat and the HLC have sought to make their orchestration activities more democratically legitimate. Enhancing participation from the developing countries and strengthening accountability are core goals of the GCA as outlined in the work program from the champions.[5] The Covid-19 pandemic in 2020 was a major setback for global climate negotiations and the GCA. However, in 2021 the secretariat and HLC embarked on an ambitious five-year work program to ramp up climate action, enhance diversity of participation, track progress, and strengthen accountability on the road to COP26 in Glasgow, culminating in the the UNFCCC Secretariat accountability and recognition framework for non party stakeholders.[6] This stands in sharp contrast to efforts before 2019, which were more fragmented and bottom-up. However, democratic legitimation is almost entirely offered to intermediaries, not to those affected (i.e., the demos). This, we argue, creates a democratic deficit that requires further attention for amelioration (Stevenson and Dryzek 2014).

## *Participation*

The HLCs and the secretariat seek to increase nonstate actor participation in the implementation of the Paris Agreement. Indeed, in 2017 the HLCs claimed that a core goal of their role was to "focus on strengthening initiatives and broadening participation by bringing on new initiatives, coalitions and actors who are committed to the implementation of action consistent with the aims of the Paris Agreement."[7] As such, the champions and the secretariat work together to orchestrate climate action by mobilizing intermediaries, with impact actors on the ground. At COP26 in Glasgow the aforementioned workplan of the HLCs – with the goal to enhance inclusion and diversity of nonstate actors in the GCA – was backed up with a number of mechanisms (discussed later).

The champions and secretariat enable participation through many different channels, the most obvious of which is the GCAP platform. NAZCA contains commitments from more than 26,309 intermediaries and 150 cooperative initiatives, that is, joint climate action by constellations of nonstate and subnational actors. The intermediary actors are cities, regions, companies, investor groups, civil society, and academic organizations. The actions include decarbonatization and adaption policies in terms of land usage, ocean and costal zones, water treatment and sustainability, building, transportation, energy, and industry. Well over 75 percent of the actors are cities or companies. To register with NAZCA, actors must enlist with the Carbon Disclosure Project, ICLEI for local governments, Climate Bonds Initiative, the Climate Group, the Global Covenant of Mayors for Climate and Energy, the Global Investor Coalition on Climate Change, or the United Nations Global Compact.

---

[5] https://unfccc.int/sites/default/files/resource/Improved%20Marrakech%20Partnership%202021-2025.pdf
[6] https://unfccc.int/sites/default/files/resource/Improved%20Marrakech%20Partnership%202021-2025.pdf
[7] https://unfccc.int/sites/default/files/gca_approach.pdf

These meta-intermediaries monitor the commitments of intermediaries, tracking their actions and reporting back to the secretariat and HLCs.

We know from previous research that this activity is heavily skewed toward the Global North (Chan et al. 2019). In 2016, around 85 percent of NAZCA participation was from the North (Galvanizing the Groundswell of Climate Actions 2015), while in 2022 it was 79 percent as outlined in a GCAP synthesis report.[8] These numbers are hard to assess, though, as Global North actions, such as those concerning industry and manufacturing, have supply chains that run through the South. However, we have seen a shift in participation post-2015. From 2015 to 2019, the total number of stakeholders rose by 60 percent and commitments to action rose by 40 percent. Regional participation has also diversified: In percentage terms Asia-Pacific stakeholders increased by 30 percent over the timeframe, while Latin America and Caribbean stakeholder participation rose by 20 percent.[9] And as concluded in the 2021 Yearbook of Climate Action, participation had increased by 20 percent since 2020 and business actors by more than 80 percent.[10] However, Western and European nonparty stakeholders dominate with almost 17,000 participants, while the African region is represented by only 600 intermediaries.

It is beyond the scope of this study to analyze individual contributions from all intermediaries. However, some general conclusions can be drawn. First, the increased participation of actors, especially from the Global South, vulnerable communities, and civil society, is still limited despite repeated calls for diversification from the HLCs and submissions of nonparty stakeholders. Intermediaries would then be covering a wider portion of the affected demos, and this should shrink democratic deficits. Second, it is unclear how stringent meta-intermediaries are in their assessment of promises as against actual contributions. What is clear, however, is that these meta-intermediaries are looking at emissions disclosures, adaptation efforts, and financing (often green bonds). They are not determining whether the intermediaries are enabling wider "stakeholder" participation (despite the frequent usage of the word "stakeholder" in secretariat and champion documents).

In some instances, this might not be overly problematic. Cities are, at least in democracies, representative of citizens that pay local taxes and through the voting of representatives. Publicly traded companies have some degree of participation from shareholders, though not stakeholders, and this might enable representation if not participation. And some organizations will be representative of their membership (for instance, an Academy of Science will have an internal structure for

---

[8] https://unfccc.int/sites/default/files/resource/GCAP%20Synthesis%20Report_Info%20as%20at%2028%20Feb%202022.pdf
[9] https://unfccc.int/sites/default/files/resource/GCA_Yearbook2018.pdf
[10] https://unfccc.int/sites/default/files/resource/Yearbook_GCA_2021.pdf

deciding leadership). However, it remains unclear from the HLCs and the secretariat whether these intermediaries are taking the people affected into consideration and, if so, how. In other words, intermediaries are enacting policies – under the gaze of meta-intermediaries directed by the orchestration of the secretariat and HLCs – with broad and deep implications. There is no mechanism in place to ensure, or even assess, whether intermediary actions have enabled democratic participation of a wider set of affected societal stakeholders.

There are of course many other mechanisms for participation, such as the Race to Zero campaign (which represents 1,049 cities, 67 regions, 5,235 businesses, 441 of the biggest investors, and 1,039 universities), which has grown exponentially since the UN Global Climate Action in New York 2019. Furthermore, the HLCs highlight the importance of regional climate week in Africa, Asia-Pacific, Latin America, and Caribbean in 2021 and 2022 as important building blocks for diversifying participation beyond Europe. Finally, the Race to Resilience campaign was launched in 2021 to mobilize action among investors, cities, and businesses to increase resilience of four billion people from vulnerable groups and communities until 2030. Our argument here is that the GCA does not have inbuilt mechanisms to determine and follow up whether actions by intermediaries are democratically legitimate and how they affect vulnerable stakeholders. This is problematic because of the nature of orchestration: It is unclear who should ensure this balanced participation. In our view, however, it is the orchestrator who directs meta-intermediaries and offers recognition to intermediaries to ensure that those on the ground are able to shape the policies that determine their lives. As the first two HLCs noted: "[W]e believe that more can be done ... to actively include in this process more representatives from national and local governments, businesses and civil society from developing countries. We intend to ensure that they are fully engaged and represented in the global climate action agenda."[11] While some steps have been made in this regard, such as the Race to Resilience campaign and "regionalization" of climate action, participation by those on the ground remains a problem.

### *Accountability and Transparency*

While the participation and representational gaps of those affected on the ground remains, accountability and transparency mechanisms have gradually improved since COP25. In 2019 the HLCs decided to improve follow-up and tracking progress of nonstate and substate climate action. Here again, we see a shift in how the secretariat and HLCs have approached these values post-Paris and in the run-up to the global stocktake of the Paris Agreement in 2023, where contributions of

---

[11] https://unfccc.int/sites/default/files/high-level-champions-climate-action-roadmap.pdf

nonstate and substate climate action will be an input alongside NDCs. While there has been recognition of the importance of tracking progress and strengthening accountability and transparency, it is unclear how deep this runs (i.e., whether this leads to substantive changes from the orchestrator) and whether this is offered to the demos or just to intermediaries.

Again, we cannot focus on all efforts (and there are many initiatives between the secretariat and the HLCs), so we look at the GCA submissions and the yearbooks of global climate action, which are framed as the main accountability mechanisms by the UNFCCC. In the lead-up to COP22 at Marrakesh, the then HLCs – Laurence Tubiana and Hakima El Haite – called for state and nonstate actors to make submissions on how best global climate action should be scaled up. Publishing their own Roadmap for Global Climate Action, the champions called for submissions in response to five key pillars of their activity: (i) How should pre-2020 ambition be managed in terms of urgency and ambition across scales and sectors? (ii) What role should the champions play in mediating between nonstate actors and state NDCs? (iii) How should nonstate actor contributions, especially through NAZCA portal, be assessed? (iv) How should high-level events both before and during COPs be organized to gain maximum exposure? (v) How should TEMs be organized in light of the global climate action agenda?[12]

These were early efforts from the HLCs and secretariat to be publicly accountable. These actors are also accountable to states as they offer support – fiscal, legal, and normative – for their roles. But as orchestrators, the secretariat and HLCs are charged with determining how best to scale-up, measure, and track the progress of climate action. This can be a more or less accountable process. As can be seen from the earlier discussion, there was an effort to have nonparty stakeholders address directly how the champions and the secretariat should reach out. While there is little sanctioning beyond naming and shaming, this submission option offered nonparty stakeholders a chance to authorize and later hold accountable the work of orchestrators.

In practice, however, the submissions were very skewed toward resourceful established actors (Bäckstrand and Kuyper 2017). The UNFCCC Secretariat asks all nonparty stakeholders to join a constituency as part of gaining observer status at COPs and Intersessionals. These are environmental NGOs, business and industry NGOs, farmers, trade unions, indigenous organizations, research organizations, local governments, women and gender organizations, and youth organizations. In 2016 there were around sixty submissions, and these were overwhelmingly from environmental NGOs and the business community (making up around half of all submissions). Similarly, in 2019 and 2020 the HLCs invited written submissions on how to improve

---

[12] https://newsroom.unfccc.int/media/658506/high-level-champions-invitation-submissions.pdf

the GCA and received around forty submissions in both these rounds.[13] This indicates that a very small number of intermediaries are engaging in written submissions and thus offering accountability to a limited number of people on the ground.

As with participation, there is very little discussion about whether these submissions actually take on board the views of different stakeholders and how. Of course, many organizations will offer this, but it is democratically beholden on the orchestrator to ensure that accountability efforts of authorization are informed not just by intermediaries but by the wider demos affected by the orchestration efforts and intermediary actions. In the run-up to 2023 global stocktake, there is very little accountability offered in terms of authorization and no ability – as far as we can tell – for affected parties to sanction the HLCs and the secretariat for their (lack of) action (Hsu et al. 2023).

Again, there may be other modes of accountability and transparency for the orchestrators. For instance, the TEPs and TEMs process offer a chance to be transparent about nonparty stakeholder activity. As the champions note, "The TEPs should draw not only on the in-session Technical Experts Meetings (TEMs) but also on the outcomes of relevant regional and thematic meetings outside of the formal sessions of the UNFCCC. Such events, with connections to the Marrakech Partnership for Global Climate Action, can enable greater participation from experts, practitioners and implementers."[14] However, the way this is disseminated lacks transparency.

Likewise, the HLCs and the secretariat have been publishing annual yearbooks on climate action under the Marrakesh Partnership since 2018. These are seen as mechanisms for accountability and transparency. Indeed, the HLCs have mobilized intermediaries to enhance transparency. For instance, the Initiative for Climate Action Transparency (ICAT) was engaged by the them to enhance the transparency of contributions to the Paris Agreement. ICAT adopts a decidedly multistakeholder partnership, working with the Children's Investment Fund Foundation; ClimateWorks Foundation; the German Federal Ministry for the Environment, Nature Conservation, and Nuclear Safety; and the Italian Ministry for the Environment, Land and Sea.

However, and as earlier, it is not clear whether this transparency reaches the demos – those on the ground affected by climate policy. These multistakeholder initiatives clearly have network benefits in terms of information-sharing and transaction costs, but they make it very hard to determine what information is transparent and to whom. At any rate, this is transparency concerning the actions of intermediaries. However, after COP25 in Madrid 2019, reporting to enhance

---

[13] https://unfccc.int/sites/default/files/resource/Marrakech_Partnership_Achievements_2019.pdf, https://unfccc.int/sites/default/files/resource/HLC-letter2020_feedback_summary.pdf
[14] https://unfccc.int/files/paris_agreement/application/pdf/marrakech_partnership_for_global_climate_action.pdf

accountability and transparency has been strengthened at the request of the parties through several reforms proposed, including: (i) The GCAP was relaunched at COP26 in Glasgow 2021 to systematically track progress and show differences between "tracked" commitments and actions; (ii) the UN Secretary-General has established a high-level expert group to develop measurement to track climate integrity and progress of nonstate actor commitments; (iii) the yearbooks for climate action, especially in 2020 and 2021, report more systematically on the progress of transnational action along various themes and sectors; (iv) during 2021 in particular, the champions submitted regular reports on the outcomes and progress of the GCA; and (v) the data partners that track commitments have produced an annual (New Climate Institute et al. 2021). This, we suggest, limits the ability of the affected demos to hold orchestrators accountable, or view transparently, the links between the orchestrator, the intermediary, and their lives.

### *Deliberation*

The final value is that of deliberation, which was of course affected by the outbreak of the Covid-19 pandemic in 2020 when negotiations were postponed for more than a year and moved to virtual format with challenges of digital gaps. While participation could come through different forms, and accountability requires authorization and sanctioning, deliberation is about the ability of the rule-makers to justify dialogically with rule-takers the decisions they are making. As with the previous two discussions, we find that the orchestrators are engaged in "summit" deliberation through the GCA events at COPs with predominantly established accredited intermediaries. Given that the link between intermediaries and the affected demos is both unchecked and likely attenuated, this is democratically problematic.

However, there are also positive developments. The HLCs have set up modes of deliberation with nonparty stakeholders. Perhaps the most central was the Talanoa Dialogue. This was set up to enable deliberative collaboration that, evidently, was scaled up by HLCs after 2020.[15] The Talanoa Dialogue involved the champions working with the COP presidents and the secretariat to allow "gender, regional and sectoral balance. Throughout the year, the champions provided guidance to ensure the participation of NPS in the Talanoa process was effective, including on how to tell impactful stories, make effective submissions to the platform and encouraging national governments and non-Party stakeholders to convene regional Talanoas."[16] The Talanoa Dialogues resulted in a UNFCCC document concerning

---

[15] https://unfccc.int/sites/default/files/resource/MP_Work_Programme_2020-2021.pdf
[16] https://unfccc.int/climate-action/marrakech-partnership/actors/meet-the-champions/previous-champions#eq-1

the COPs in 2017.[17] This document noted that the HLCs should continue the work of the facilitative dialogue at COP21. The facilitative, as well as the later Talanoa, dialogue requires individuals and organizations to commit to climate shifts with respect to NDCs. This operated as a pre-2020 stocktake. It was guided by generally deliberative ideals: nonconfrontational, empathy/trust building, collective good-building, and so on.

This was echoed in the TEMs. Herein TEMs were suggested as a way that orchestrators could enlist intermediaries for their goals. TEMs and TEPs are enacted throughout the year, now virtually. They cover how land use, food chains, and forestry might matter for climate change. The HLCs do interact with the secretariat about this.[18] However, it is not cohesive and perhaps needs more deliberative quality in terms of those actually affected. Deliberation was limited during 2020, but the previously mentioned regional climate weeks in 2021 were intended to enhance deliberation, partnership, and collaboration between states and nonparty stakeholders and vulnerable communities on the ground. At COP26 in Glasgow both the HLCs and the UN Secretary-General participated in a series of events related to GCA.

Overall, we think that the orchestrators – the UNFCCC Secretariat and the HLCs – could be doing more. It is clear that mechanisms are diversifying, but not that varied positions are making their mark. Looking at actual citizen engagement in terms of participation, accountability/transparency, and deliberation exposes some major shortcomings.

## 8.6 Conclusions

In this chapter, we have argued that the UNFCCC Secretariat works in close collaboration with the HLCs and also the UN Secretary-General to orchestrate nonstate and substate climate action to increase ambition in the forthcoming global stocktake of the Paris Agreement in 2023. This could be seen in many ways as a joint initiative, as states have empowered the champions but asked them to work alongside the secretariat. Likewise, there are several comanaged online portals, such as GCAP/NAZCA, which deepen this relationship. The orchestrator trio – the champions, the secretariat, and the UN Secretary-General – increasingly stress the importance of diverse participation of vulnerable communities across the Global South and values of equity, resilience, and just transition in the GCA. While questions of how effective is orchestration are predominant in the academic and policy literatures, we ask a different question: Is orchestration democratically legitimate?

---

[17] https://unfccc.int/resource/docs/2017/cop23/eng/l13.pdf
[18] https://unfccc.int/resource/climateaction2020/media/1308/unfccc_spm_2018.pdf

To answer this, we suggest focusing on democratic values – participation, accountability/transparency, and deliberation. These values tap into different models of democracy and help give expression to the notion that individuals should have a say over how their lives are governed. As such, we have claimed that those affected should have a say in how their lives are directed and constrained.

Our analysis is deliberately circumspect. We are not claiming that the UNFCCC Secretariat and the HLCs are – or are not – democratically legitimate. There are a wide variety of mechanisms such as high-level events at COPs/Intersessionals, NAZCA, the yearbooks of climate action the UNFCCC Secretariat Recognition and Accountability Framework, and TEMs, that substantiate deepened engagement on how to reduce the "participatory" gap between the North and the South, business and civil society, in the Marrakech Partnership. We propose that the democratic legitimacy of these efforts should be given equal attention to effectiveness and be evaluated more systematically in line with our framework. That is, we should ask how the orchestrator, using the intermediaries, remains – or fails – to be democratically legitimated by those actually affected.

At this stage, it appears that democratic legitimacy is weak. Participation after seven years since the birth of the Marrakech Partnership is heavily skewed toward actors in the Global North, and there is no check on how much say stakeholders actually have in the position of their representatives. Accountability is also low, in the sense that those affected cannot authorize or sanction orchestrators. However, in 2021, reporting and tracking of contributions of nonstate commitments were substantively improved through a revamped GCAP. Transparency has also been strengthened by these initiatives from the orchestrators, but without accountability, this is a weak value. Finally, deliberation is limited to the high-level summit format occurring mostly between established intermediaries of businesses, cities, and investors, rather than actually engaging with those on the ground affected by climate hazards. However, the "regionalization" of climate action UNFCCC Secretariat Recognition and Accountability Framework, through regional climate weeks can potentially increase both deliberation and participation from national and local stakeholders.

Future research on the democratic legitimacy of orchestrated global climate action should focus on two streams. First, what is the precise relationship between the secretariat, the HLCs, and the UN Secretary-General? They operate in similar spaces, but the nature of the relationship is understudied. It seems clear that the HLCs were set up on the margins of the secretariat structure, but with much bidirectional cooperation needed to fulfill each other's goals. Both the UN Global Climate Action Summit 2019 and Climate Ambition Summit in 2023 hosted by the UN Secretary-General meant a boost to climate action and has strengthened collaboration between different orchestrators. Second, there is a question as to whether individuals and citizens implicated in orchestrated initiatives are able to democratically legitimize their intermediaries. We should then examine whether the affected individuals – through cities, regions, firms, or other organizations – have a chance to shape the polices that affect their lives. In turn, we should study whether orchestrators take this on board in their decision-making (i.e., in their relationship with intermediaries).

Ultimately, the orchestration of nonstate climate action might increase effectiveness and ambition, as current transparency efforts seem to suggest. But asking whether the democratic legitimacy of orchestration is always good requires thinking about people on the ground. These people should decide how their governing rules are decided. As such, probing the democratic legitimacy of orchestration might help ensure that relationships between the orchestrator, meta-intermediary, and intermediary are clear, as well as probing whether uptake on the ground is enacted.

## References

Abbott, K. W., Genschel, P., Snidal, D., and Zangl, B. (eds.) (2015). *International Organizations as Orchestrators*, Cambridge: Cambridge University Press.

Abbott, K. W, Genschel P., Snidal. D., and Zangl, B. (2016). Two Logics of Indirect Governance: Delegation and Orchestration, *British Journal of Political Science* 46 (4): 719–729.

Abizadeh, A. (2012). On the Demos and Its Kin: Nationalism, Democracy, and the Boundary Problem, *American Political Science Review* 106 (4): 867–882.

Bäckstrand, K. (2008). Accountability of Networked Climate Governance: The Rise of Transnational Climate Partnerships, *Global Environmental Politics* 8 (3): 74–104.

Bäckstrand, K. and Kuyper, J. W. (2017). The Democratic Legitimacy of Orchestration: The UNFCCC, Non-state Actors, and Transnational Climate Governance, *Environmental Politics* 26 (4): 764–788.

Chan, S. and Amling, W. (2019). Does Orchestration in the Global Climate Action Agenda Effectively Prioritize and Mobilize Transnational Climate Adaptation Action? *International Environmental Agreements* 19: 429–446.

Chan, S., Falkner, R., Goldberg, M., & Van Asselt, H. (2018). Effective and Geographically Balanced? An Output-Based Assessment of Non-State Climate Actions, *Climate Policy* 18 (1): 24–35.

Chan, S., Boran, I., and van Asselt, H., et al. (2019). Promises and Risks of Nonstate Action in Climate and Sustainability Governance, *WIREs Climate Change* 10 (3): 572.

Dingwerth, K. (2014). Global Democracy and the Democratic Minimum: Why a Procedural Account Alone Is Insufficient, *European Journal of International Relations* 20 (4): 1124–1147.

Falkner, R. (2016). The Paris Agreement and the New Logic of International Climate Politics, *International Affairs* 92 (5): 1107–1125.

Galvanizing the Groundswell of Climate Actions (2015). *Lima-Paris Action Agenda Independent Assessment Report*. www.climategroundswell.org/blog-test/lpaa/report

Goodin, R. E. (2007). Enfranchising All Affected Interests, and Its Alternatives, *Philosophy & Public Affairs* 35 (1): 40–68.

Gordon, D. J. and Johnson, C. A. (2017). The Orchestration of Global Urban Climate Governance: Conducting Power in the Post-Paris Climate Regime, *Environmental Politics* 26 (4): 694–714.

Graham, E. R. (2017). The Institutional Design of Funding Rules at International Organizations: Explaining the Transformation in Financing the United Nations, *European Journal of International Relations* 23 (2): 365–390.

Grant, R. and Keohane, R. O. (2005). Accountability and Abuses of Power in World Politics, *American Political Science Review* 99 (31): 29–44.

Habermas, J. (1996). *Between Facts and Norms*, Cambridge, MA: MIT Press.

Hale, T. (2016). "All Hands on Deck": The Paris Agreement and Nonstate Climate Action, *Global Environmental Politics* 16 (3): 12–22.

Hale, T. and Roger, C. (2014). Orchestration and Transnational Governance, *Review of International Organization* 9 (1): 59–82.

Hale, T. N., Chan, S., Hsu, A., Clapper, A., Elliott, C., Faria, P., …, & Widerberg, O. (2021). Sub-and Non-State Climate Action: A Framework to Assess Progress, Implementation and Impact, *Climate Policy*, 21 (3): 406–420.

Hickmann, T. and Elsässer, J. (2020). New Alliances in Global Environmental Governance: How Intergovernmental Treaty Secretariats Interact with Non-state Actors to Address Transboundary Environmental Problems, *International Environmental Agreements: Politics, Law and Economics* 20 (3): 459–481.

Hickmann, T., Widerberg, O., Lederer, M., and Pattberg, P. (2021). The United Nations Framework Convention on Climate Change Secretariat as an Orchestrator in Global Climate Policymaking, *International Review of Administrative Sciences* 87 (1): 21–38.

Hsu, A., Moffat, A. S., Weinfurter, A. J., and Schwartz, J. D. (2015). Towards a New Climate Diplomacy, *Nature Climate Change* 5 (6): 501–503.

Hsu, A., Chan, S., Roelfsema, M., Schletz, M., Kuramochi, T., Smit, S., and Deneault, A. (2023). From Drumbeating to Marching: Assessing Non-state and Subnational Climate Action Using Data, *One Earth* 6 (9), 1077–1081.

Koenig-Archibugi, M. (2017). How to Diagnose Democratic Deficits in Global Politics: The Use of the "All-Affected Principle," *International Theory* 9 (2): 171–202.

Kuyper, J. W. (2014). Global Democratization and International Regime Complexity, *European Journal of International Relations* 20 (3): 620–646.

Macdonald, T. (2008). *Global Stakeholder Democracy: Power and Representation beyond Liberal States*, Oxford: Oxford University Press.

McKenna, C, et al. (2022). Integrity Matters: Net Zero Commitments by Businesses, Financial Institutions, Cities and Regions. Report from the United Nations' High-Level Expert Group on the Net Zero Emissions Commitments of Non-State Entities.

Nasiritousi, N. and Grimm, J. (2022). Governing toward Decarbonization: The Legitimacy of National Orchestration. *Environmental Policy and Governance* 32 (5): 411–425.

New Climate Institute, Data-Driven EnviroLab, Utrecht University, et al. (2021). *Global Climate Action from Cities, Regions and Businesses: Taking Stock of the Impact of Individual Actors and Cooperative Initiatives on Global Greenhouse Gas Emissions*, Cologne: New Climate Institute. https://newclimate.org/sites/default/files/2021/06/NewClimate_GCC_June21_2.pdf.

Peixoto, T. (2013). The Uncertain Relationship between Open Data and Accountability: A Response to Yu and Robinson, *UCLA Law Review* 60: 200–248.

Stevenson, H. and Dryzek, J. S. (2014). *Democratizing Global Climate Governance*, Cambridge: Cambridge University Press.

UN News (2021). *COP26: Enough of "Treating Nature Like a Toilet"*. https://news.un.org/en/story/2021/11/1104542.

UNFCCC Secretariat (2023) Recognition and Accountability Framework for Non-Party Stakeholder Climate Action, June 2023.

# 9

# The Administrative Embeddedness of International Environmental Secretariats

*Toward a Global Administrative Space?*

BARBARA SAERBECK, HELGE JÖRGENS, ALEXANDRA GORITZ, JOHANNES SCHUSTER, MAREIKE WELL, AND NINA KOLLECK[*]

## 9.1 Introduction

Today's global governance system is characterized by institutional complexity, bottom-up and top-down elements, and a multiplicity of actors and levels. Public administrations are generally seen as an important element of this global governance architecture (Bauer, Knill, and Eckhard 2017). Kingsbury and Stewart (2005: 17) even argue that "much of global governance can be understood and analyzed as administrative action: rule making, administrative adjudication between competing interests, and other forms of regulatory and administrative decisions and management." Coining the term "global administrative law," they and others call "for the recognition of a global administrative space in which international and transnational administrative bodies interact in complex ways" (Wessel and Wouters 2007: 281) and in which states are no longer the single determinant but rather one among many.

The concept of a global administrative space relates to the institutional structure that underlies processes of global policymaking, namely the emergence of administrative structures beyond the territory of the nation-state (Kingsbury and Stewart 2005). However, we still lack knowledge about the embeddedness, role, and position of environmental bureaucracies in their respective networks and how they interact with other types of actors. Only recently have scholars begun to study the interaction between state and nonstate actors and environmental bureaucracies within the architectures of global environment governance (see, e.g., Saerbeck et al. 2020; Wit et al. 2020). Applying the notion of an administrative space to the global environmental governance regime promises to be a fruitful endeavor as it is believed that not just the state signatories of a convention contribute to

---

[*] The German Research Foundation under Grants JO 1142/1-1 and KO 4997/1-1 supported this work. We would like to thank Kyra Ksinzyk, Vanessa Höhne, Flávia Rabello and Susanne Helm for their assistance in preparing the data on which the study is based on as well as for their valuable comments.

processes of environmental multilateral decision-making. Rather, it is assumed that environmental bureaucracies and state and nonstate actors have formed complex networks, thereby strengthening the bond between once disconnected entities. As such, it is argued that, similar to the European administrative space, one needs "to stop thinking in terms of hierarchical layers of competence separated by the subsidiarity principle and start thinking, instead, of a networking arrangement, with all levels of governance shaping, proposing, implementing and monitoring policy together" (Prodi 2000 in Martens 2006: 126).

This contribution seeks to deepen our understanding of the global environmental governance regime, and in particular the role of environmental bureaucracies within it. We argue that state and nonstate actors as well as environmental bureaucracies operating on various levels interact with one another within the global environmental governance regime. Furthermore, we argue that international public administrations play a central role not only in the global environmental governance regime but also in the global environmental administrative space. Building on an original dataset of issue-specific cooperation and information flows among organizations active in the global climate and the biodiversity regimes, we test our arguments by studying whether environmental bureaucracies, state organizations, and nonstate organizations interact horizontally and vertically with one another.

We assess our argument by means of social network analysis. This allows us to detect the diverse interactions that environmental bureaucracies cultivate with one another as well as with state and nonstate actors. Based on an original data set derived from a large-N survey among organizations in two fields of global environmental governance, our social network analysis maps networks of policy-specific communication and cooperation among diverse actor groups. This approach enables us to assess the position, the embeddedness, and the potential role of specific actors within these networks. Moreover, we can draw conclusions about the relationships between various actor types within the same negotiations.

Our study speaks to the literature on global environmental governance architectures (Aldy and Stavins 2007; Biermann et al. 2009; Keohane and Victor 2011). The literature on the global climate governance regime has focused mainly on the interaction between negotiation parties and nonparty actors (see, e.g., Nasiritousi and Linnér 2016; Nasiritousi, Hjerpe, and Buhr 2014, 2016; Tallberg et al. 2013), thereby somewhat neglecting the link between administrations and state and nonstate actors (for a recent exception see Biermann and Kim 2020). Our approach focuses on the bureaucratic side of these governance arrangements and how they interact with others, a focus that we consider to be of great importance. For example, scholars of international public administration study their agency and influential role in multilateral negotiations by inquiring whether, how, and to which degree they exert influence on international policymaking (see, e.g., Bauer 2006,

2009; Bauer, Andresen, and Biermann 2012; Bauer and Ege 2017; Busch 2009; Eckhard and Ege 2016; Jinnah 2011, 2014; Johnson 2016; Saerbeck et al. 2020; Tallberg et al. 2013; Well et al. 2020). These scholars find that international bureaucracies partially act beyond the mandate state actors grant them, trying to mobilize support to advance their own proposals and to build momentum for multilateral agreements (Abbott and Snidal 2010; Jörgens et al. 2017; Saerbeck et al. 2020). They can be powerful actors that wield (independent) influence in global policymaking by controlling information and the ability to transform this information into knowledge – that is, to structure perceptions (Barnett and Finnemore 2004). International bureaucracies exert influence, inter alia, through the use of their central position in actor networks, their privileged access to information, their professional authority, and technical expertise (Bauer and Ege 2016; Jinnah 2014; Jörgens et al. 2017; Widerberg and van Laerhoven 2014).

The next section reviews concepts of inter- and transnational administrative spaces to study the phenomenon of administrative structures and state- and non-state actors' networks. This allows us to formulate first expectations about the characteristics of a potentially emerging global environmental administrative space. The following section builds on an original dataset derived from a large-N survey among organizations operating in the fields of global climate and biodiversity governance to empirically map networks of policy-specific communication and cooperation. This allows us to assess the global environmental governance structure as well as the position that administrative organizations occupy within this regime. We discuss our findings in the conclusion, in which we also outline avenues for future research.

## 9.2 Concepts of International and Transnational Administrative Spaces

The international and transnational administrative spaces are relatively new concepts in the public administration and international relations literature. They were systematically dealt with for the first time in the context of European integration research. In the following, we first look at the characteristics of the so-called European administrative space before we review the concepts of global or transnational administrative structures that are not bound to the polity of the European Union.

### *The European and Global Administrative Spaces*

The European administrative space is a nonhierarchical order of closely intertwined operational and decision-making levels combined with a major structural variability. A first wave of research on the European administrative space focused mainly on the convergence of (national) administrative systems "on a common

European model" (Olsen 2003: 506), in which a "public administration operates and is managed on the basis of European principles, rules and regulations uniformly enforced in the relevant territory" (Olsen 2003: 508). Closely related to more general notions of European integration or Europeanization of national political systems, the Europeanization of national public administrations was seen as "a new pattern of European integration that complements regulatory integration" (Trondal and Peters 2015: 79). The emergence of a European administrative space was thought to be a cross-national convergence of national administrations.

Arguing that the convergence of national administrative systems was at best inconsistent and incomplete (Knill 2001; Olsen 2003), a second wave of research focused on the multilevel character of European public administration. Departing from a predominant focus on the substantial attributes of administrative organizations – such as size and expertise of staff, financial resources, or formal mandates and competencies – to including relational attributes of public administrations led to a reconceptualization of the European administrative space as network-based rather than state-centric. From this perspective, a European administrative space emerges through the intensification of relationships between (integrated) administrative units at different levels of government, that is, a vertical pooling of administrative resources from different levels of government within particular policy domains or issue areas (Benz 2015; Hofmann 2008). The European administrative space is seen as "a space in which European, national and sub-national administrations and interested parties act together in agenda setting, rule-making and implementation" (Hofmann 2008: 670). According to Heidbreder (2011: 710), it is best understood "in procedural terms as a network marked by 'functional unity', 'organizational separation' and 'procedural co-operation.'"

Overall, the prevailing notion of a European administrative space can be described as "a common European administrative infrastructure for the joint formulation and execution of public policy" (Trondal and Peters 2015: 79) with established links to relevant nonstate actors within a given issue area or policy domain. Its main features are (i) an interest of public administrations at different levels of government as political actors, (ii) a focus on their relationships and interactions with other bureaucracies as well as with other (non)state actors, and (iii) a governance perspective that is interested in processes of formulating and implementing political programs within the European multilevel polity. Research in this tradition is rooted simultaneously in the subdisciplines of public administration, public policy, European studies, and international relations.

Trondal and Peters (2013, 2015) moreover identify three analytical dimensions that characterize the European administrative space – institutional independence, integration, and co-optation. The first dimension, institutional independence, "involves the institutionalization of some level of independent administrative capacity" at the international level, which the authors characterize as "relatively permanent and separate institutions that are able to act relatively independently from [national] governments" (Trondal and Peters 2015: 80). The second dimension, integration, "entails ... the inter-institutional integration of administrative structures" at the global level. Finally, the third dimension, co-optation, means that "there is a mutual process of integration" of domestic agencies, regional administrative structures such as the institutions of the European Union, international bureaucracies, and nongovernmental organizations (NGOs) at all levels of government that are involved in the exercise of administrative tasks (Trondal and Peters 2015: 80).

In contrast to the concept of European administrative spaces, the term "global administrative space" is not frequently used in the fields of public administration, international relations and international law. It figures most prominently in the work of Kingsbury, Krisch, and Stewart (2005: 25), who see the recognition of a distinct global administrative space as a way to overcome "the classical dichotomy between an administrative space in national polities on the one hand and inter-state coordination in global governance on the other." They relate the concept of the global administrative space to the emergence of administrative structures beyond the nation-state. Kingsbury and Stewart (2008: 3–4) characterize this space as being "populated by several distinct types of regulatory administrative institutions and various types of entities that are the subject of regulation, including not only states but firms, NGOs and individuals."[1] While their notion of a global administrative space shows a considerable degree of overlap with that of transnational governmental networks, it differs from the latter in that it defines the global administrative space in functional rather than formal terms. In their understanding, the global administrative space is restricted not to formal bureaucratic organizations (or their individual members) but to those organizations (and their individual members) that actually perform administrative functions at all levels of government.[2]

## *Transgovernmental Networks and Multilevel Governance*

A number of approaches describe and analyze the emergence of administrative structures beyond the European Union. A very early field of study was what Nye and Keohane (1971b: 331) termed "transnational relations." Transnational relations

---

[1] Kingsbury and Stewart (2008: 3–4) distinguish between five groups of actors in the global administrative space: (i) "formal intergovernmental organizations," especially their "internal organs of an administrative character," (ii) "intergovernmental networks of national regulatory officials," (iii) "hybrid intergovernmental-private bodies, composed of both public and private actors," (iv) "private bodies exercising public governance functions," and (v) "domestic administrative agencies whose regulatory decisions significantly affect other countries or their citizens."

[2] Both definitions include the relations and interactions of administrative organizations with their respective target audiences.

are defined as "contacts, coalitions, and interactions across state boundaries that are not controlled by the central foreign policy organs of governments." While not specifically focusing on the interaction of administrative actors at different levels of government, Nye and Keohane explicitly acknowledge that public administrations at the (sub)national level may act in partial autonomy from their own governments when interacting with state or nonstate actors in other countries or at the international level. They observe that "subunits of governments may ... have distinct foreign policies which are not all filtered through the top leadership and which do not fit into a unitary actor model" (Nye and Keohane 1971a: 731). At the same time, international secretariats may seek transgovernmental actors "as potential allies" (Nye and Keohane 1971a: 748) and can be expected "to form explicit or implicit coalitions with sub-units of governments as well as with nongovernmental organizations having similar interests" (Keohane and Nye 1974: 52).

In a state-of-the-art review on transnational relations, Nölke (2016) characterized transnational politics as a space in which a wide range of organizations, including businesses, NGOs, research institutes, national ministries, agencies, subnational governments, and international public administrations, interact and form transnational policy networks. Slaughter (2004) argues that transnational networks of government officials have substituted traditional diplomacy in many policy areas. Building on Keohane and Nye (1974) and Slaughter (2004), Hale and Held (2011: 16) define transgovernmental networks as more or less formalized fora that "bring 'domestic' government officials together with their peers around specific issues, often regulatory in nature."

Multilevel governance approaches moreover cover linkages between the public and private sector more generally and between state and supranational authority specifically. They describe the complex distribution and linkages as well as the blurred boundaries of competencies and responsibilities between state and nonstate activities at different levels: "Multi-Level Governance posits that decision-making authority is not monopolized by the governments of the member states but is diffused to different levels of decision-making, the sub-national, national and supranational levels" (Marks 1993: 392). The multilevel governance approach "focuses on the change in form of the exercise of political power and the new forms of cooperation and coordination that transcend 'hierarchy' (in the sense of central control) and 'market' (in the sense of spontaneous, unplanned self-organization)" (Huster 2008: 56). Whereas multilevel refers to the growing independence of the system from governments, the term "Governance" is a reference to the growing interdependence of state and nonstate actors (Bache and Flinders 2004: 2–3). Various forms of governance at different levels of decision-making are connected to form an overall composition of "Governance by, with, and without Government" (Zürn 1998: 166–167).

## 9.3 A Global Environmental Administrative Space?

Transnational relations as well as multilevel governance approaches direct the scholarly focus to the linkages between the public and private sphere on the international level. They conceptualize international governance processes as a space in which state and nonstate actors form complex policy networks. Studies of global climate governance echo this notion and describe the global environmental governance structure as highly dynamic relationships within and between different levels of governance and government (Biermann et al. 2009; Saerbeck et al. 2020). For example, the climate regime that is based on the United Nations Framework Convention on Climate Change's (UNFCCC) Paris Agreement has been described as a "hybrid system that combines bottom-up with top-down elements" (Falkner 2016: 21), which emphasizes the role and importance of issue-specific initiatives carried out by a diverse set of actors (Fuhr and Hickmann 2016; Jänicke and Quitzow 2017; Pattberg and Stripple 2008).[3] European/global administrative space approaches suggest that the existence of network-based administrative structures lies beyond the nation-state. They point to the multilevel character of public administrations.

Our argument is that a global environmental administrative space is currently emerging within the global environmental governance regime through the intensification of relationships between (integrated) administrative units at different levels of government. We also believe that environmental bureaucracies have formed bonds not only with one another but also with state and nonstate actors operating at different levels in the global environmental regime. Studies on international public administrations focus on interorganizational cooperation and issue-specific information flows (see, e.g., Bauer, Knill, and Eckhard 2017; Saerbeck et al. 2020). Well et al. (2020), moreover, showed that expertise and information are more important tools for international public administrations than rules and formal powers. While the formal mandates and legal competencies of international public administrations are rather limited when compared to national bureaucracies, their strategic use of expertise, ideas, and procedural knowledge combined with their mostly central position in issue-specific information flows forms the basis of their impact on global policy outputs (Busch and Liese 2017). International public administrations actively shape their organizational environment by setting up and forming structures of multilevel administration and by creating informal alliances with nonstate actors at all levels of government. International bureaucracies then typically occupy a central position in "their" domain-specific organizational

---

[3] Its structure intends to facilitate (inter)action, learning, and diffusion of best practices between a wide variety of actors operating across levels and sectors through the provision of multiple access points (Jänicke 2017; Jörgensen and Wagner 2017; Ostrom 2010).

environment, especially within domain-specific information flows (Benz, Corcaci, and Doser 2017; Jörgens et al. 2017; Saerbeck et al. 2020; see also Chapter 4). As such, we expect international public administrations to be prominent actors within the global environmental administrative space as well as the global environmental governance regime.

H1: A global administrative space has emerged within the global environmental governance regime, in which environmental bureaucracies of all levels interact with each another.
H2: The global environmental administrative space comprises networks between environmental bureaucracies and state and nonstate actors.
H3: International public administrations play a prominent role in the global environmental administrative space.

## 9.4 Mapping the Global Environmental Administrative Space

In his article on the development of a European administrative space, Olsen (2003: 506) asks, "How can we recognize an EAS [European administrative space] if one has emerged?" The same question applies to this chapter: How can we define, operationalize, and measure a potential global environmental administrative space? In this section, we will propose social network analysis as a method to respond to this challenge.

We believe that a global environmental administrative space can be best observed through a systematic empirical analysis of policy-related information flows and cooperation between different kinds of actors that are directly or indirectly involved in global environmental governance. Social network analysis focuses on social relations between actors and the resulting network structures, instead of actors' individual attributes (Jörgens, Kolleck, and Saerbeck 2016). It allows us to map the issue-specific network of organizations operating within a given policy domain to identify relationships and types of interactions among them and, as such, to study the interaction patterns of state and nonstate actors as well as environmental bureaucracies.

Our analysis is based on data that we collected via a large-N online questionnaire between September 2015 and March 2016. In this questionnaire we approached a wide variety of state and nonstate actors operating at different levels in the global climate and biodiversity regimes.[4] We received 471 (sometimes partial) responses for the UNFCCC and 561 for the Convention on Biological

---

[4] For the two surveys, we identified the respondents through lists of the Conference of the Parties participants in previous years. We then extended the number of respondents based on the snowball principle and data provided in open questions.

Diversity (CBD). The questionnaire included two questions on the relationships between actors. The first question asked about cooperation among different actors ("Which organizations did you cooperate closely with regarding topics discussed under the UNFCCC/CBD during the last 12 months?") and the second about information provision ("Which organizations did you receive trustworthy information from during the last 12 months?").[5] Since both questions provide information on relationships relevant to the emergence of a global environmental governance regime and a global environmental administrative space, we combined the answers. This gives us an idea of whether interaction takes place among environmental bureaucracies themselves or between environmental bureaucracies and state and nonstate actors across different environmental issue areas (climate and biodiversity).

To measure the embeddedness of individual organizations in the combined network, we calculate different measures of centrality. The higher an organization's centrality value, the higher its embeddedness in the global environmental governance regime and its global environmental administrative space (see, e.g., Hanneman and Riddle 2011). First, degree centrality measures how many relationships an actor has within a given network. In our case, the degree centrality measures how often an actor is named as a source of policy-relevant information or a cooperation partner and how often an actor is the one who named others. It is a measure for reputation and general visibility in a network. Second, eigenvector centrality indicates the prominence of an actor in a network by measuring whether it is linked to other important actors. An actor's eigenvector centrality is high only if the contacts also have a high eigenvector centrality. Such an actor may have only a few, but very important, relations. Finally, betweenness centrality measures how often an actor is positioned on the closest path between any other two actors within the network. In an information exchange network, for example, a high betweenness centrality enables an actor to alter the information that is being exchanged between different actors.

The next sections describe the global policy network that evolved around the UNFCCC and the CBD. The edges represent either instances of interorganizational cooperation or instances of communication where one organization

---

[5] Respondents who did not respond to this survey item were spread equally across the different categories of participants. We left out invalid responses, commonly resulting from the impossibility of identifying the mentioned organization due to misspelled acronyms or other reasons. The responses to the two questions moreover allowed only for a maximum of six answers. The combined network therefore does not represent the totality of existing cooperative or communicative links between the organizations in the network, but only those that are most highly valued by the survey's respondents. This is also the reason why we did not calculate any measures to describe the overall network structure, such as network density, reciprocity, transitivity, or average path length (see Hanneman and Riddle 2011), as any measure would be strongly biased and underestimate the coherence of the network.

receives trustworthy information related to the UNFCCC or the CBD from another organization. We distinguish between six actor groups (governments, international organizations [IOs], NGOs, research institutes, private businesses including banks, and others), which enables us to learn more about the relative centrality of different types of organizations. In the first step, we provide network graphs and tables with centrality values for the top thirty organizations to develop an initial understanding of the global environmental governance regime complex. Next, we draw our attention to the embeddedness of environmental bureaucracies as well as their interactions with state and nonstate actors within that regime to determine the characteristics of the global environmental administrative space.[6]

## *The Global Environmental Governance Regime*

This section visualizes the current global environmental governance regime to gain a better understanding of the interaction taking place between state and nonstate actors and environmental bureaucracies. Figures 9.1 and 9.2 visualize the combined UNFCCC and CBD network. While the colors of the nodes in Figure 9.1 represent actor groups, the colors in Figure 9.2 indicate which of the two UN conventions an organization can be primarily attributed to. Table 9.1 lists the thirty organizations with the highest centrality values in the combined network.

Figure 9.1 shows the current global environmental governance regime. From a structural perspective, it is particularly interesting that the network consists of one main component of connected actors, while only a few actors are not involved in any sort of interaction with this component. Despite tendencies for polycentricism (Jordan et al. 2018), there are core actors to the global environmental governance regime that are closely connected. No systematic structures of group formations in relation to actor type can be observed, indicating that all actor types engage in interactions with other types of stakeholders. When looking at the position of specific actors (see Table 9.1), we see that IOs are at the core of the current global environmental governance system. Interestingly, these are not only the United Nations Environment Programme (UNEP) and the United Nations Development Programme (UNDP), two IOs that play leadership roles in environmental and development policy, but also the two convention secretariats.

---

[6] Although we find Kingsbury and Stewart's (2008) approach of including bureaucratic organizations and organizations who actually perform administrative functions at all levels of government in the conceptualization of a global administrative space interesting, we refrain from using their definition of a bureaucratic actor due to restrictions caused by our methodological approach.

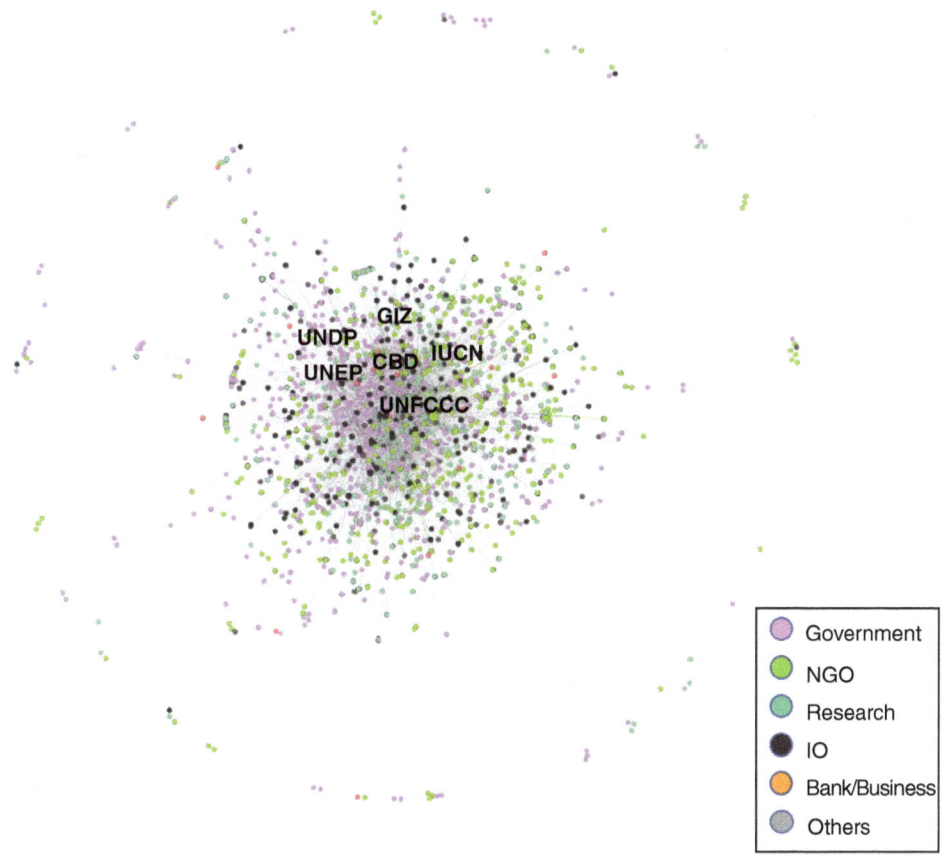

Figure 9.1 The combined CBD and UNFCCC network by actor groups (node size refers to degree centrality, and node color refers to actor group)

Figure 9.2 and Table 9.1 suggest that an institutional structure has evolved that is present in different issue-specific networks of global environmental governance. This structure comprises international (e.g., UNEP, UNDP, European Commission, the Food and Agriculture Organization [FAO]), governmental (different national environmental ministries and agencies), nongovernmental (e.g., WWF, CAN) and research organizations (e.g., WRI, CGIAR). Some organizations, such as the IUCN and the IPCC, are themselves compound organizations with traits of an IO and NGO (IUCN) or a research organization (IPCC), respectively.[7] However, the results suggest that the global environmental governance

---

[7] UNEP = United Nations Environment Program; UNDP = United Nations Development Program; WWF = World Wide Fund for Nature; CAN = Climate Action Network; WRI = World Resources Institute; IUCN = International Union for the Conservation of Nature; IPCC = Intergovernmental Panel on Climate Change.

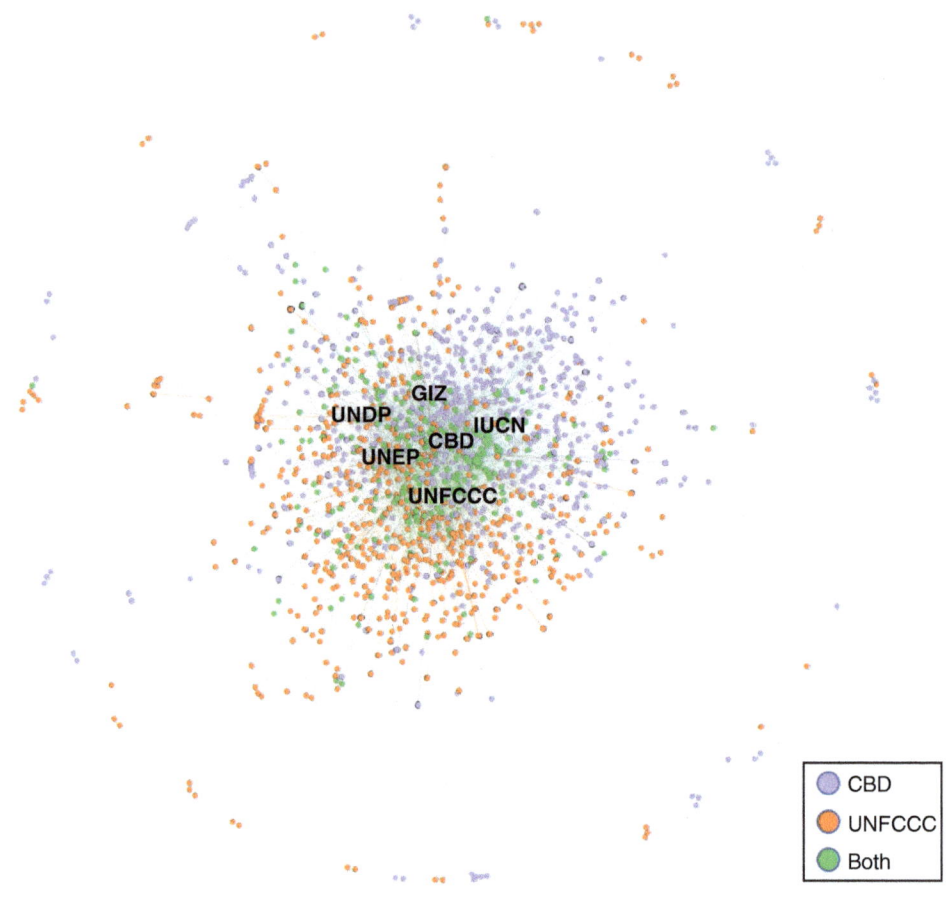

Figure 9.2 The combined CBD and UNFCCC network by UN conventions (node size refers to degree centrality, and node color refers to convention)

regime is mostly dominated by IOs, NGOs, and governmental actors, while only a few research organizations and businesses can be found among the most central actors.

The results suggest an embeddedness of environmental bureaucracies within the global environmental governance structure. The high centrality scores of international public administrations (according to all three centrality measures presented in Table 9.1) indicate that they occupy a central position in their respective treaty networks. Figure 9.2 and the betweenness centrality scores in Table 9.1 also show that in particular the CBD Secretariat occupies a very central position in the combined network that could be an indicator of a bridge function between the climate (orange) and the biodiversity (blue) regime.

Table 9.1 Top thirty organizations with the highest centrality values in the combined CBD and UNFCCC network

| | Degree centrality | | | | Betweenness centrality | | | | Eigenvector centrality | | | |
|---|---|---|---|---|---|---|---|---|---|---|---|---|
| | Organization | Type | Value | Con. | Organization | Type | Value | Con. | Organization | Type | Value | |
| 1 | UNEP | IO | 133 | Both | CBD | IO | 281,093 | CBD | UNDP | IO | 1 | cbd |
| 2 | UNDP | IO | 132 | CBD | UNEP | IO | 240,335 | Both | CBD | IO | 0.96 | cbd |
| 3 | CBD | IO | 119 | CBD | UNDP | IO | 232,583 | CBD | UNEP | IO | 0.93 | both |
| 4 | IUCN | IO | 106 | CBD | UNFCCC | IO | 191,920 | Both | UNFCCC | IO | 0.85 | both |
| 5 | GIZ, Germany | Gov. | 102 | CBD | IUCN | IO | 173,717 | Both | IUCN | IO | 0.82 | cbd |
| 6 | UNFCCC | IO | 95 | Both | GIZ, Germany | Gov. | 159,406 | CBD | GIZ | Gov. | 0.81 | cbd |
| 7 | WWF | NGO | 88 | Both | WWF | NGO | 123,716 | Both | WWF | NGO | 0.71 | both |
| 8 | MoEFCC, India | Gov. | 69 | Both | MoEFCC, India | Gov. | 109,146 | Both | CGIAR | Res. | 0.51 | cbd |
| 9 | CGIAR | Res. | 61 | CBD | CGIAR | Res. | 84,386 | CBD | FAO | IO | 0.51 | both |
| 10 | FAO | IO | 51 | Both | FAO | IO | 68,475 | Both | BMUB | Gov. | 0.42 | both |
| 11 | EU Commission | IO | 45 | Both | CAN | NGO | 56,380 | UNFCCC | MoEFCC, India | Gov. | 0.40 | both |
| 12 | BMUB, Germany | Gov. | 40 | Both | BMUB, Germany | Gov. | 48,863 | Both | WRI | Res. | 0.40 | both |
| 13 | CI | NGO | 37 | Both | UNESCO | IO | 48,334 | Both | EU | IO | 0.37 | both |
| 14 | CAN | NGO | 36 | UNFCCC | EU Commission | IO | 47,703 | Both | EU Commission | IO | 0.36 | both |
| 15 | GEF | IO | 35 | Both | CI | NGO | 43,787 | Both | World Bank | IO | 0.35 | both |
| 16 | IPCC | IO | 34 | Both | SPREP | IO | 37,538 | Both | MoE, Peru | Gov. | 0.35 | both |
| 17 | WRI | Res. | 34 | Both | IPCC | IO | 34,461 | Both | Go4BioDiv | NGO | 0.34 | cbd |
| 18 | EU Council | IO | 33 | CBD | DoECC, Canada | Gov. | 33,127 | Both | CI | NGO | 0.32 | both |
| 19 | EU | IO | 33 | Both | EU Council | IO | 31,376 | CBD | CAN | NGO | 0.31 | unfccc |
| 20 | DETEC, Switzerland | Gov. | 32 | Both | BirdLife | NGO | 30,137 | CBD | MoNRE, Thailand | Gov. | 0.31 | both |

Table 9.1 (cont.)

| | Degree centrality | | | | Betweenness centrality | | | | Eigenvector centrality | | | |
|---|---|---|---|---|---|---|---|---|---|---|---|---|
| | Organization | Type | Value | Con. | Organization | Type | Value | Con. | Organization | Type | Value | |
| 21 | World Bank | IO | 31 | Both | DETEC, Switzerland | Gov. | 30,101 | Both | ICIMOD | IO | 0.29 | both |
| 22 | CIFOR | Res. | 31 | Both | DEA | Gov. | 29,285 | Both | IPCC | IO | 0.29 | both |
| 23 | BirdLife | NGO | 29 | CBD | WRI | Res. | 29,278 | Both | OECD | IO | 0.29 | cbd |
| 24 | MoE, Peru | Gov. | 28 | CBD | MoE, Peru | Gov. | 29,129 | Both | MoEW, Bolivia | Gov. | 0.28 | both |
| 25 | Go4BioDiv | NGO | 28 | Both | World Bank | IO | 29,111 | Both | MINAE, Costa Rica | Gov. | 0.28 | both |
| 26 | MoE, Japan | Gov. | 26 | Both | DOEE, Australia | Gov. | 28,780 | Both | IETA | Bus. | 0.27 | unfccc |
| 27 | OECD | IO | 26 | CBD | BNHS | NGO | 28,757 | CBD | EIB | IO | 0.27 | both |
| 28 | SPREP | IO | 24 | Both | Go4BioDiv | NGO | 28,734 | CBD | GHMC, India | Gov. | 0.27 | both |
| 29 | ICIMOD | IO | 24 | Both | CIFOR | Res. | 26,300 | Both | GEF | IO | 0.26 | both |
| 30 | BNHS | NGO | 23 | CBD | MoEW, Bolivia | Gov. | 26,156 | Both | ASEAN | IO | 0.25 | both |

## A Global Environmental Administrative Space

To answer the question whether a global administrative space has emerged within the global environmental governance regime, in which environmental bureaucracies of all levels interact with one another as well as with state and nonstate actors, we focus on their interactions. At first, we look at the interactions of environmental bureaucracies with one another. For this purpose, we reduce the network to interactions of government actors and IOs. We assume that the answers given by our survey respondents that named IOs and government actors mostly refer to the administrative parts of these organizations.[8]

The colors of the nodes indicate the convention the administrative actors can be attributed to. The structure of the graph in Figure 9.3 suggests that the environmental bureaucracies not only engage in cooperation and exchange of information within the scope of their respective convention but they also interact with public administrations from other environmental issue areas. Again, this applies particularly to the two convention secretariats. Table 9.2 lists the thirty environmental bureaucracies with the highest centrality values. Although no local actors can be found among the thirty most central actors, the presence of bureaucracies that belong to both IOs and national agencies and ministries indicates that vertical interaction patterns emerge in addition to the horizontal interactions observed from the network graph. These results serve as a first indicator for the integration of administrative structures and thus the existence of a global environmental administrative space.

To further investigate the existence of a global environmental administrative space, we study the interactions between environmental bureaucracies and state and nonstate actors and the position of international public administration within this network. Figures 9.4 and 9.5 visualize the information exchange and cooperation of environmental bureaucracies with state and nonstate actors. In contrast to the network presented in Figures 9.1 and 9.2, we created this network by using only relations that involved administrative actors, either as the source or the target of interaction; hence, these figures can be interpreted as egocentric networks of the administrative actors involved. In this way, we can analyze co-optation, the mutual process of integration of domestic administrations, regional administrative institutions such as the European Union, international bureaucracies, and nongovernmental organizations at different levels of government (Trondal and Peters 2015: 80). Again, the colors of the nodes in Figure 9.4 represent actor groups and the colors in Figure 9.5 indicate to which of the two UN conventions an actor belongs to. We then calculated the centrality measures for the actors involved in this network in

---

[8] See, for example, Well et al. (Chapter 4) who point to the need to treat IO and their bureaucracies as actors in their own right, as autonomous and consequential actors and not as instruments of nation-states.

Figure 9.3 Network of environmental bureaucracies (node size refers to degree centrality, and node color refers to convention)

order to identify particularly central actors. Table 9.3 lists the thirty organizations with the highest centrality values in the global environmental administrative space.

Figure 9.4 shows that the global administrative space comprises IOs, governmental administrations, NGOs, research organizations, and businesses. As could be seen already in the overall network, the administrations associated with IOs are mainly positioned in the center of the graph, while other stakeholders are evenly distributed. At the same time, the structure indicates that the global environmental administrative space comprises various state and nonstate actors that engage in cooperation and exchange of information with environmental bureaucracies. Similar to the previous findings, Table 9.3 shows that the actors with the highest centrality values belong to IOs, while research organizations and businesses are underrepresented. The high number of governmental actors among the most central actors again indicates that interactions emerge not only between various actors but also between different levels of governance.

Table 9.2 The thirty environmental administrative actors with the highest centrality values in the global environmental administrative space

| | Degree centrality | | | | Betweenness centrality | | | | Eigenvector centrality | | | |
|---|---|---|---|---|---|---|---|---|---|---|---|---|
| | Organization | Type | Value | Con. | Organization | Type | Value | Con. | Organization | Type | Value | Con. |
| 1 | UNDP | IO | 91 | CBD | UNEP | IO | 55,906 | Both | UNDP | IO | 1 | cbd |
| 2 | UNEP | IO | 83 | Both | UNDP | IO | 55,311 | CBD | UNEP | IO | 0.93 | both |
| 3 | GIZ, Germany | Gov. | 60 | CBD | CBD | IO | 46,844 | CBD | CBD | IO | 0.80 | cbd |
| 4 | CBD | IO | 56 | CBD | GIZ, Germany | Gov. | 32,210 | CBD | GIZ, Germany | Gov. | 0.69 | cbd |
| 5 | IUCN | IO | 52 | CBD | IUCN | IO | 26,669 | CBD | UNFCCC | IO | 0.67 | both |
| 6 | UNFCCC | IO | 43 | Both | UNFCCC | IO | 25,268 | Both | IUCN | IO | 0.67 | cbd |
| 7 | FAO | IO | 35 | Both | FAO | IO | 21,647 | Both | FAO | IO | 0.53 | both |
| 8 | EU Council | IO | 32 | CBD | EU Council | IO | 16,468 | CBD | EU Commission | IO | 0.38 | both |
| 9 | EU Commission | IO | 28 | Both | DoECC, Canada | Gov. | 13,851 | Both | EU | IO | 0.38 | both |
| 10 | EU | IO | 25 | Both | SPREP | IO | 13,224 | Both | BMUB, Germany | Gov. | 0.34 | both |
| 11 | MoEFCC, India | Gov. | 25 | Both | MoEFCC, India | Gov. | 12,428 | Both | OECD | IO | 0.32 | cbd |
| 12 | DETEC, Switzerland | Gov. | 25 | Both | EU Commission | IO | 11,672 | Both | MoNRE, Thailand | Gov. | 0.32 | both |
| 13 | GEF | IO | 20 | Both | UNESCO | IO | 10,932 | Both | MoE, Peru | Gov. | 0.32 | both |
| 14 | DoECC, Canada | Gov. | 20 | Both | DETEC, Switzerland | Gov. | 8,390 | Both | MoEW, Bolivia | Gov. | 0.31 | both |
| 15 | ICIMOD | IO | 20 | Both | EU | IO | 7,658 | Both | EIB | IO | 0.30 | both |
| 16 | SPREP | IO | 20 | Both | MoEE, Sweden | Gov. | 7,636 | Both | World Bank | IO | 0.28 | both |

Table 9.2 (cont.)

| | | Degree centrality | | | Betweenness centrality | | | | Eigenvector centrality | | | |
|---|---|---|---|---|---|---|---|---|---|---|---|---|
| | Organization | Type | Value | Con. | Organization | Type | Value | Con. | Organization | Type | Value | Con. |
| 17 | BMUB, Germany | Gov. | 19 | Both | BMUB | Gov. | 7,525 | Both | ICIMOD | IO | 0.28 | both |
| 18 | World Bank | IO | 19 | Both | MoNRE, Malaysia | Gov. | 7,372 | Both | NAMA Facility | IO | 0.28 | unfccc |
| 19 | OECD | IO | 18 | CBD | ICIMOD | IO | 7,212 | Both | IPCC | IO | 0.28 | both |
| 20 | MoEW, Bolivia | Gov. | 16 | Both | World Bank | IO | 7,210 | Both | NEAA, Netherlands | Gov. | 0.27 | both |
| 21 | MoE, Sweden | Gov. | 16 | Both | BMLFUW, Austria | Gov. | 6,084 | Both | MINAE, Costa Rica | Gov. | 0.27 | both |
| 22 | UNESCO | IO | 16 | Both | SEMARNAT, Mexico | Gov. | 6,001 | Both | MoE, Moldova | Gov. | 0.27 | both |
| 23 | MoNRE, Malaysia | Gov. | 15 | Both | COMIFAC | IO | 5,196 | Both | DENR, Philippines | Gov. | 0.25 | both |
| 24 | IPCC | IO | 15 | Both | MoCE, Norway | Gov. | 5,068 | Both | SEMARNAT, Mexico | Gov. | 0.25 | both |
| 25 | COMIFAC | IO | 14 | Both | NEPA, Afghanistan | Gov. | 4,900 | UNFCCC | ASEAN | IO | 0.24 | both |
| 26 | BMLFUW, Austria | Gov. | 14 | Both | GWP | IO | 4,850 | UNFCCC | DETEC, Switzerland | Gov. | 0.24 | both |
| 27 | SEMARNAT, Mexico | Gov. | 14 | Both | OECD | IO | 4,692 | CBD | GEF | IO | 0.24 | both |
| 28 | MoC, Norway | Gov. | 14 | Both | South Africa | Gov. | 4,690 | Both | UNESCO | IO | 0.24 | both |
| 29 | South Africa | Gov. | 14 | Both | MoEW, Bolivia | Gov. | 4,522 | Both | BMLFUW, Austria | Gov. | 0.23 | both |
| 30 | NAMA Facility | IO | 14 | UNFCCC | DEA, South Africa | Gov. | 4,451 | Both | MoCE, Norway | Gov. | 0.23 | both |

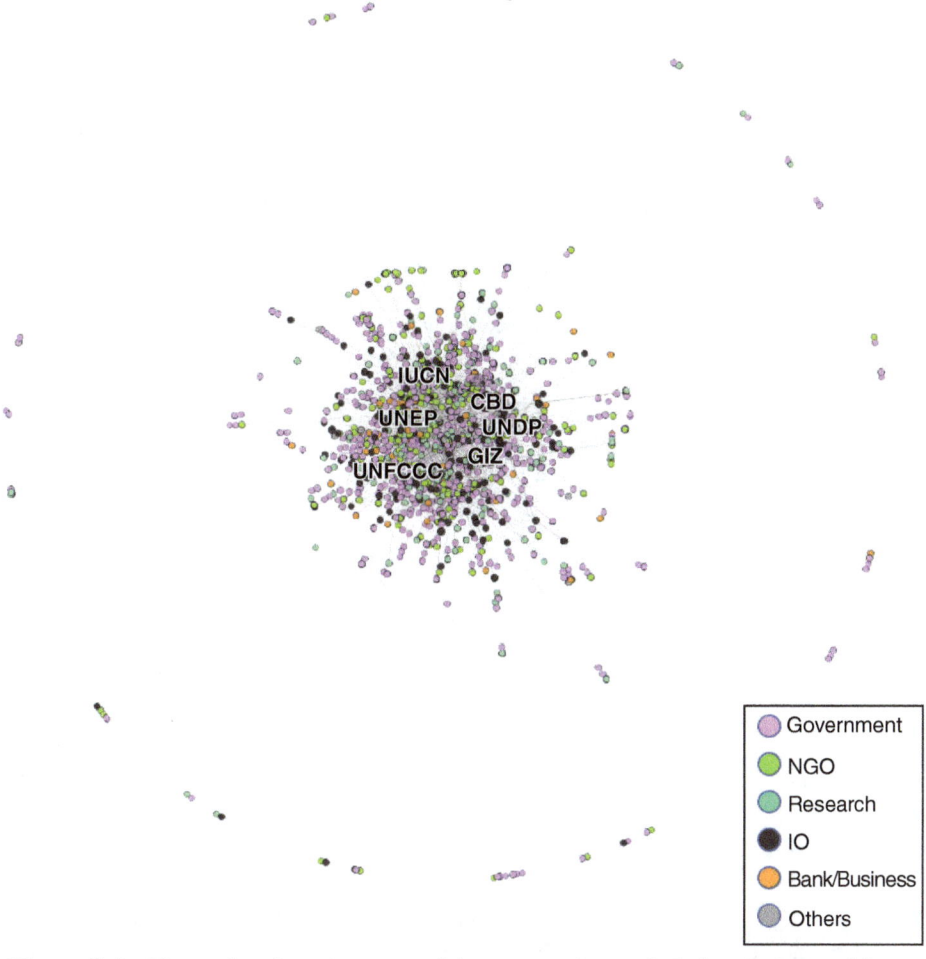

Figure 9.4 Network of environmental bureaucracies and their relations with state and nonstate actors by actor group (node size refers to degree centrality, and node color refers to actor group)

## 9.5 Conclusions

This chapter studied the characteristics of the global administrative space and the embeddedness of environmental bureaucracies within that space. We applied concepts of inter- and transnational relations (e.g., transgovernmental networks, multilevel governance approaches, and the European/global administrative space) and used social network analysis. The latter allowed us to describe the current global environmental governance regime and to systematically examine the environmental bureaucrat's relations. Building on an original dataset on issue-specific cooperation and information flows among organizations active in the global climate and

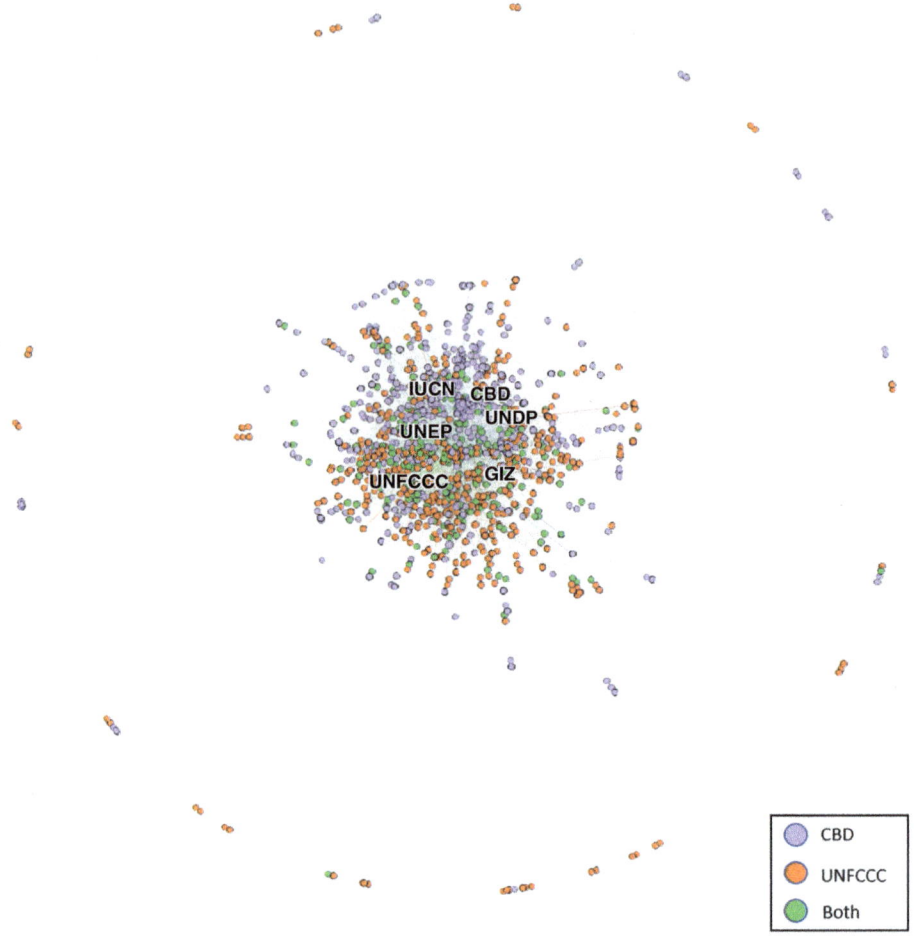

Figure 9.5 Network of environmental bureaucracies and their relations with state and nonstate actors by UN convention (node size refers to degree centrality, and node color refers to convention)

the biodiversity regimes, we find that environmental bureaucracies interact with one another as well as with state and nonstate actors within the global environmental governance regime. They have succeeded in forming complex networks of relations stretching from the local and national to the international level, constituting an emerging global environmental administrative space.

We moreover discover that environmental bureaucracies, mostly international public administrations, occupy central positions within the global environmental governance regime, even bridging the two environmental treaty conventions under study. Their high centrality scores indicate that they are engaged in cooperation and information exchange with organizations that are more strongly involved in

Table 9.3 The thirty organizations with the highest centrality values in the global environmental administrative space

| | Degree centrality | | | | Betweenness centrality | | | | Eigenvector centrality | | | |
|---|---|---|---|---|---|---|---|---|---|---|---|---|
| | Organization | Type | Value | Con. | Organization | Type | Value | Con. | Organization | Type | Value | Con. |
| 1 | UNEP | IO | 133 | Both | CBD | IO | 200,585 | CBD | UNDP | IO | 1 | cbd |
| 2 | UNDP | IO | 132 | CBD | UNEP | IO | 188,596 | Both | UNEP | IO | 0.95 | both |
| 3 | CBD | IO | 119 | CBD | UNDP | IO | 172,434 | CBD | CBD | IO | 0.95 | cbd |
| 4 | IUCN | IO | 106 | CBD | UNFCCC | IO | 125,348 | Both | UNFCCC | IO | 0.82 | both |
| 5 | GIZ, Germany | Gov. | 102 | CBD | IUCN | IO | 121,972 | CBD | IUCN | IO | 0.80 | cbd |
| 6 | UNFCCC | IO | 95 | Both | GIZ, Germany | Gov. | 116,954 | CBD | GIZ, Germany | Gov. | 0.80 | cbd |
| 7 | MoEFCC, India | Gov. | 69 | Both | MoEFCC, India | Gov. | 69,369 | Both | FAO | IO | 0.50 | both |
| 8 | FAO | IO | 51 | Both | FAO | IO | 53,776 | Both | WWF | NGO | 0.49 | both |
| 9 | WWF | NGO | 48 | Both | WWF | NGO | 42,061 | Both | CGIAR | Res. | 0.41 | cbd |
| 10 | EU Commission | IO | 45 | Both | UNESCO | IO | 39,124 | Both | BMUB, Germany | Gov. | 0.41 | both |
| 11 | BMUB, Germany | Gov. | 40 | Both | EU Commission | IO | 37,960 | Both | MoEFCC, India | Gov. | 0.37 | both |
| 12 | CGIAR | Res. | 37 | CBD | CGIAR | Res. | 37,930 | CBD | EU Commission | IO | 0.37 | both |
| 13 | GEF | IO | 35 | Both | BMUB, Germany | Gov. | 35,163 | Both | EU | IO | 0.35 | both |
| 14 | IPCC | IO | 34 | Both | SPREP | IO | 30,195 | Both | World Bank | IO | 0.34 | both |
| 15 | EU | IO | 33 | Both | DoECC, Canada | Gov. | 28,902 | Both | MoE, Peru | Gov. | 0.34 | both |
| 16 | EU Council | IO | 33 | CBD | CI | NGO | 28,200 | Both | MoNRE, Thailand | Gov. | 0.31 | both |

Table 9.3 (cont.)

| | Degree centrality | | | | Betweenness centrality | | | | Eigenvector centrality | | | |
|---|---|---|---|---|---|---|---|---|---|---|---|---|
| | Organization | Type | Value | Con. | Organization | Type | Value | Con. | Organization | Type | Value | Con. |
| 17 | DETEC, Switzerland | Gov. | 32 | Both | EU Council | IO | 25,723 | CBD | CI | NGO | 0.30 | both |
| 18 | World Bank | IO | 31 | Both | DETEC, Switzerland | Gov. | 25,516 | Both | OECD | IO | 0.29 | cbd |
| 19 | CI | NGO | 29 | Both | World Bank | IO | 24,458 | Both | ICIMOD | IO | 0.29 | both |
| 20 | MoE, Peru | Gov. | 28 | Both | IPCC | IO | 24,411 | Both | IPCC | IO | 0.29 | both |
| 21 | MoE, Japan | Gov. | 26 | Both | DOEE, Australia | Gov. | 24,346 | Both | MoEW, Bolivia | Gov. | 0.28 | both |
| 22 | OECD | IO | 26 | CBD | MoEW, Bolivia | Gov. | 21,004 | Both | Go4BioDiv | NGO | 0.28 | cbd |
| 23 | ICIMOD | IO | 24 | Both | MoE, Peru | Gov. | 20,880 | Both | MINAE, Costa Rica | Gov. | 0.27 | both |
| 24 | SPREP | IO | 24 | Both | DEA, South Africa | Gov. | 20,473 | Both | EIB | IO | 0.27 | both |
| 25 | UNESCO | IO | 22 | Both | EU | IO | 19,480 | Both | IETA | Bus. | 0.27 | unfccc |
| 26 | SEMARNAT, Mexico | Gov. | 22 | Both | MoNRE, Malaysia | Gov. | 19,367 | Both | GEF | IO | 0.26 | both |
| 27 | UBA, Germany | Gov. | 21 | UNFCCC | Climate Analytics | Res. | 18,998 | UNFCCC | GHMC, India | Gov. | 0.25 | both |
| 28 | IEA | IO | 21 | UNFCCC | ICIMOD | IO | 17,496 | Both | UNESCO | IO | 0.24 | both |
| 29 | IOC | IO | 21 | Both | GEF | IO | 17,391 | Both | NEAA, Netherlands | Gov. | 0.24 | both |
| 30 | MoEW, Bolivia | Gov. | 21 | Both | MoE, Brazil | Gov. | 17,298 | Both | WRI | Res. | 0.24 | both |

the negotiation and implementation of the other convention, thereby attempting to connect broader policy discourses with specific negotiation items. This may be a sign of (formal or informal) autonomy that they have acquired vis-à-vis state actors. It would be worthwhile to take into account the challenges that may arise in principal–agent relations,[9] as our results highlight the importance of a research agenda that focuses on potential autonomy of environmental bureaucracies as well as their functionality, structure, and legitimacy. International public administrations might aim to gain autonomy from their principals and seek influence in environmental policy processes, for example, by defining and framing problems, exchanging information about best practices, and proposing solutions that are potentially affecting the decision-makers at different levels of government.

The high number of governmental actors furthermore indicates that interactions emerge not only between various actors but also beyond different levels of governance. A multiplicity of sometimes overlapping environmental institutions have been detected, including numerous environmental treaty bodies such as the climate and biodiversity secretariats as well as various IOs that formally belong to other policy domains but whose tasks are in part immediately relevant for global environmental issues. These organizations include, among others, the World Bank and the FAO. Finally, we find that some organizations, such as the IUCN and the IPCC, are themselves compound organizations with traits of an IO, performing administrative tasks, and an NGO (IUCN) or a research organization (IPCC), respectively. These findings direct the attention to the administrative tasks that are being performed by diverse state and nonstate actors at different levels in a given policy domain. Taking our results as a starting point, future research could investigate whether these interactions also lead to processes of integration among administrative actors across different levels of government and to the co-optation of nongovernmental or semigovernmental actors within a common global environmental administrative structure. Kingsbury, Krisch, and Stewart (2005: 22–23), for example, argue that at the international level private organizations, which they refer to as "hybrid intergovernmental-private administration," fulfill functions similar to those of public administrations at the national level and propose to study such bodies "as part of global administration." Further studies need to analyze the integration among administrative actors across different levels of government and of co-optation of nongovernmental or semigovernmental actors within a common global environmental administrative structure.

---

[9] The principal–agent approach tries to explain how contractual partners pursue their commitments despite an asymmetric distribution of information and diverging interests, and under the premises of utility-maximizing or opportunistic actors. A major risk is shirking by the agent, also known as "agency drift": Administrations may develop an institutional self-interest and exploit the information asymmetry vis-à-vis the principal resulting from unclear negotiation levels spread over numerous hierarchical levels to pursue their goals.

## References

Abbott, K. W. and Snidal, D. (2010). International Regulation without International Government: Improving IO Performance through Orchestration, *Review of International Organizations* 5 (3): 315–344.

Aldy, J. E. and Stavins, R. N. (eds.) (2007). *Architectures for Agreement: Addressing Global Climate Change in the Post-Kyoto-World*, Cambridge: Cambridge University Press.

Bache, I. and Flinders, M. (eds.) (2004). *Multi-level Governance*, Oxford: Oxford University Press.

Barnett, M. and Finnemore, M. (2004). *Rules for the World: International Organizations in Global Politics*, Ithaca, NY: Cornell University Press.

Bauer, M. W. and Ege, J. (2016). Bureaucratic Autonomy of International Organizations' Secretariats, *Journal of European Public Policy* 23 (7): 1019–1037.

Bauer, M. W. and Ege, J. (2017). A Matter of Will and Action: The Bureaucratic Autonomy of International Public Administrations. In M. W. Bauer, C. Knill, and S. Eckhard (eds.), *International Bureaucracy: Challenges and Lessons for Public Administration Research*. Basingstoke: Palgrave Macmillan: 13–41.

Bauer, M. W., Knill, C., and Eckhard, S. (2017). International Public Administration: A New Type of Bureaucracy? Lessons and Challenges for Public Administration Research. In M. W. Bauer, C. Knill, and S. Eckhard (eds.), *International Bureaucracy: Challenges and Lessons for Public Administration Research*, Basingstoke: Palgrave Macmillan, 179–198.

Bauer, S. (2006). Does Bureaucracy Really Matter? The Authority of Intergovernmental Treaty Secretariats in Global Environmental Politics, *Global Environmental Politics* 6 (1): 23–49.

Bauer, S. (2009). The Desertification Secretariat: A Castle Made of Sand. In F. Biermann and B. Siebenhüner (eds.), *Managers of Global Change: The Influence of International Environmental Bureaucracies*, Cambridge, MA: MIT Press, 293–317.

Bauer, S., Andresen, S., and Biermann, F. (2012). International Bureaucracies. In F. Biermann and P. Pattberg (eds.), *Earth System Governance: Global Environmental Governance Reconsidered*, Cambridge, MA: MIT Press, 27–44.

Benz, A. (2015). European Public Administration as a Multilevel Administration: A Conceptual Framework. In M. W. Bauer and J. Trondal (eds.), *The Palgrave Handbook of the European Administrative System*, Basingstoke: Palgrave Macmillan, 31–47.

Benz, A., Corcaci, A., and Doser, J. W. (2017). Multilevel Administration in International and National Contexts. In M. W. Bauer, C. Knill, and S. Eckhard (eds.), *International Bureaucracy: Challenges and Lessons for Public Administration Research*, Basingstoke: Palgrave Macmillan, 151–178.

Biermann, F. and Kim, R. E. (eds.) (2020). *Architectures of Earth System Governance: Institutional Complexity and Structural Transformation*, Cambridge: Cambridge University Press.

Biermann, F., Pattberg, P., van Asselt, H., and Zelli, F. (2009). The Fragmentation of Global Governance Architectures: A Framework for Analysis, *Global Environmental Politics* 9 (4): 14–40.

Busch, P.-O. (2009). The Climate Secretariat: Making a Living in a Straitjacket. In F. Biermann and B. Siebenhüner (eds.), *Managers of Global Change: The Influence of International Environmental Bureaucracies*, Cambridge, MA: MIT Press, 245–264.

Busch, P.-O. and Liese, A. (2017). The Authority of International Public Administrations. In M. W. Bauer, C. Knill, and S. Eckhard (eds.), *International Bureaucracy: Challenges and Lessons for Public Administration Research*, Basingstoke: Palgrave Macmillan, 97–122.

Eckhard, S. and Ege, J. (2016). International Bureaucracies and Their Influence on Policy-Making: A Review of Empirical Evidence, *Journal of European Public Policy* 23 (7): 960–978.

Falkner, R. (2016). The Paris Agreement and the New Logic of International Climate Politics, *International Affairs* 92 (5): 1107–1125.

Fuhr, H. and Hickmann, T. (2016). Transnationale Klimainitiativen und die internationalen Klimaverhandlungen, *Zeitschrift für Umweltpolitik und Umweltrecht* 39: 88–94.

Hale, T. and Held, D. (eds.) (2011). *Handbook of Transnational Governance: Institutions and Innovations*, Cambridge: Polity.

Hanneman, R. A. and Riddle, M. (2011). Concepts and Measures for Basic Network Analysis. In J. Scott and P. J. Carrington (eds.), *The SAGE Handbook of Social Network Analysis*, London: SAGE Publications, 340–369.

Heidbreder, E. G. (2011). Structuring the European Administrative Space: Policy Instruments of Multi-level Administration, *Journal of European Public Policy* 18 (5): 709–727.

Hofmann, H. C. H. (2008). Mapping the European Administrative Space, *West European Politics* 31 (4): 662–676.

Huster, S. (2008). *Europapolitik aus dem Ausschuss*, Wiesbaden: VS Verlag für Sozialwissenschaften.

Jänicke, M. (2017). The Multi-level System of Global Climate Governance: The Model and Its Current State, *Environmental Policy and Governance* 27 (2): 108–121.

Jänicke, M. and Quitzow, R. (2017). Multi-level Reinforcement in European Climate and Energy Governance: Mobilizing Economic Interests at the Sub-national Levels, *Environmental Policy and Governance* 27 (2): 122–136.

Jinnah, S. (2011). Marketing Linkages: Secretariat Governance of the Climate-Biodiversity Interface, *Global Environmental Politics* 11 (3): 23–43.

Jinnah, S. (2014). *Post-Treaty Politics: Secretariat Influence in Global Environmental Governance*, Cambridge, MA: MIT Press.

Johnson, T. (2016). Cooperation, Co-optation, Competition, Conflict: International Bureaucracies and Non-governmental Organizations in an Interdependent World, *Review of International Political Economy*, 1–31. https://doi.org/10.1080/09692290.2016.1217902

Jordan, A., Huitema, D., van Asselt, H., and Forster, J. (eds.) (2018). *Governing Climate Change: Polycentricity in Action?*, Cambridge: Cambridge University Press.

Jörgens, H., Kolleck, N., and Saerbeck, B. (2016). Exploring the Hidden Influence of International Treaty Secretariats: Using Social Network Analysis to Analyse the Twitter Debate on the "Lima Work Programme on Gender," *Journal of European Public Policy* 23 (7): 979–998.

Jörgens, H., Kolleck, N., Saerbeck, B., and Well, M. (2017). Orchestrating (Bio-)Diversity: The Secretariat of the Convention of Biological Diversity as an Attention-Seeking Bureaucracy. In M. W. Bauer, C. Knill, and S. Eckhard (eds.), *International Bureaucracy: Challenges and Lessons for Public Administration Research*, Basingstoke: Palgrave Macmillan, 73–95.

Jörgensen, K. and Wagner, C. (2017). Low Carbon Governance in Multi-level Structures. EU-India Relations on Energy and Climate, *Environmental Policy and Governance* 27 (2): 137–148.

Keohane, R. O. and Nye, J. S. (1974). Transgovernmental Relations and International Organizations, *World Politics* 27 (1): 39–62.

Keohane, R. O. and Victor, D. G. (2011). The Regime Complex for Climate Change, *Perspectives on Politics* 9 (1): 7–23.

Kingsbury, B. and Stewart, R. B. (2008). Legitimacy and Accountability in Global Regulatory Governance: The Emerging Global Administrative Law and the Design and Operation of Administrative Tribunals of International Organizations. In

K. Papanikolaou (ed.), *International Administrative Tribunals in a Changing World*, London: Esperia Publications.

Kingsbury, B., Krisch, N., and Stewart, R. B. (2005). The Emergence of Global Administrative Law, *Law and Contemporary Problems* (3–4): 15–61.

Knill, C. (2001). *The Europeanisation of National Administrations: Patterns of Institutional Change and Persistence, Change and Persistence*, Cambridge: Cambridge University Press.

Marks, G. (1993). Structural Policy and Multilevel Governance in the EC. In A. Cafruny and G. Rosenthal (eds.), *The State of the European Community. Vol. 2: The Maastricht Debates and Beyond*, Boulder, CO: Lynne Rienner, 391–410.

Martens, M. (2006). National Regulators between Union and Governments: A Study of the EU's Environmental Policy Network IMPEL. In M. Egeberg (ed.), *Multilevel Union Administration: The Transformation of Executive Politics*, Basingstoke: Palgrave Macmillan, 124–142.

Nasiritousi, N. and Linnér, B.-O. (2016). Open or Closed Meetings? Explaining Nonparty Actor Involvement in the International Climate Change Negotiations, *International Environmental Agreements: Politics, Law and Economics* 16 (1): 127–144.

Nasiritousi, N., Hjerpe, M., and Buhr, K. (2014). Pluralising Climate Change Solutions? Views Held and Voiced by Participants at the International Climate Change Negotiations, *Ecological Economics* 105: 177–184.

Nasiritousi, N., Hjerpe, M., and Linnér, B.-O. (2016). The Roles of Non-state Actors in Climate Change Governance: Understanding Agency through Governance Profiles, *International Environmental Agreements: Politics, Law and Economics* 16(1): 109–126.

Nölke, A. (2016). International Relations and Transnational Politics. In H. Keman and J. J. Woldendorp (eds.), *Handbook of Research Methods and Applications in Political Science*, Cheltenham: Edward Elgar, 169–183.

Nye, J. S. and Keohane, R. O. (1971a). Transnational Relations and World Politics: A Conclusion, *International Organization* 25 (3): 721–748.

Nye, J. S. and Keohane, R. O. (1971b). Transnational Relations and World Politics: An Introduction, *International Organization* 25 (3): 329–349.

Olsen, J. P. (2003). Towards a European Administrative Space?, *Journal of European Public Policy* 10 (4): 506–531.

Ostrom, E. (2010). Polycentric Systems for Coping with Collective Action and Global Environmental Change, *Global Environmental Change* 20 (4): 550–557.

Pattberg, P. and Stripple, J. (2008). Beyond the Public and Private Divide: Remapping Transnational Climate Governance in the 21st Century, *International Environmental Agreements: Politics, Law and Economics* 8 (4): 367–388.

Saerbeck, B., Well, M., Jörgens, H., Goritz, A., and Kolleck, N. (2020). Brokering Climate Action: The UNFCCC Secretariat between Parties and Nonparty Stakeholders, *Global Environmental Politics* 20 (2): 105–127.

Slaughter, A.-M. (2004). *A New World Order*, Princeton: Princeton University Press.

Tallberg, J., Sommerer, T., Squatrito, T., and Jonsson, C. (2013). *The Opening Up of International Organizations: Transnational Access in Global Governance*, Cambridge: Cambridge University Press. https://doi.org/10.1017/CBO9781107325135

Trondal, J. and Peters, B. G. (2013). The Rise of European Administrative Space: Lessons Learned, *Journal of European Public Policy* 20 (2): 295–307.

Trondal, J. and Peters, B. G. (2015). A Conceptual Account of the European Administrative Space. In M. W. Bauer and J. Trondal (eds.), *The Palgrave Handbook of the European Administrative System*, Basingstoke: Palgrave Macmillan, 79–92.

Well, M., Saerbeck, B., Jörgens, H., and Kolleck, N. (2020). Between Mandate and Motivation: Bureaucratic Behavior in Global Climate Governance, *Global Governance: A Review of Multilateralism and International Organizations* 26 (1): 99–120.

Wessel, R. A. and Wouters, J. (2007). The Phenomenon of Multilevel Regulation: Interactions between Global, EU and National Regulatory Spheres, *International Organizations Law Review* 4 (2): 259–291.

Widerberg, O. and van Laerhoven, F. (2014). Measuring the Autonomous Influence of an International Bureaucracy: The Division for Sustainable Development, *International Environmental Agreements: Politics, Law and Economics* 14 (4): 303–327.

Wit, D. de, Ostovar, A. L., Bauer, S., and Jinnah, S. (2020). International Bureaucracies. In F. Biermann and R. E. Kim (eds.), *Architectures of Earth System Governance: Institutional Complexity and Structural Transformation*, Cambridge: Cambridge University Press, 57–74.

Zürn, M. (1998). *Democratic Governance beyond the Nation State?*, IIS-Arbeitspapier 12/78, Bremen.

# 10

# Reflections on the Role of International Public Administrations in the Anthropocene

FRANK BIERMANN

## 10.1 Introduction

John Bolton, former US ambassador to the United Nations and Donald Trump's national security advisor, once quipped, "There's no such thing as the United Nations. ... If the U.N. secretary building in New York lost 10 stories, it wouldn't make a bit of difference" (Bolton 1994). This sentiment is widely shared in conservative circles around the world, even if rarely articulated as bluntly. Yet also in academia and the study of international relations, disregard of an autonomous political influence of intergovernmental organizations and the United Nations is widely spread. In many study programs, world politics is still defined as a system shaped by states, with only a marginal role for international organizations as independent agents in global policy processes.

This volume joins the growing chorus of those who break with this traditional approach and who argue for more serious academic engagement with international organizations and the public administrations at their core. Within the larger debate on international public administrations, this volume makes a crucial intervention in its theoretical focus on bureaucratic autonomy and agency by strengthening and further developing the research program on international bureaucracies that has started many years ago. In this concluding chapter I reflect on the key contributions of this book, considering both earlier work and the new challenges for international public administrations in the Anthropocene.

## 10.2 *Managers of Global Change*: A Reassessment

My own interest in the study of international public administrations dates back to the late 1990s, when I began to study the deficiencies of the UN system in global environmental governance. In 2000 I developed with Bernd Siebenhüner a major research program on international environmental bureaucracies, which concluded with the publication of *Managers of Global Change: The Influence of International Environmental Bureaucracies* (Biermann and Siebenhüner 2009).

This project contributed to a broader theoretical turn toward the study of international organizations in international relations research. When we conceptualized *Managers of Global Change*, international relations research was dominated by neoinstitutionalist and neorealist theoretical strands along with the emergent critique of constructivism and international political economy (see overview by Bauer et al. 2009). None of these approaches, at that time, gave much prominence to international bureaucracies and to the civil servants working in these organizations. After 2000, however, several research projects had begun to address this gap, and international bureaucracies became a more widely studied phenomenon (e.g., Barnett and Finnemore 2004; Hawkins et al. 2006; Johnson and Urpelainen 2014). *Managers of Global Change* has been a part of this conceptual turn, with a focus on global environmental politics.

*Managers of Global Change* tried to make several conceptual contributions. One was our differentiation between normative and administrative structures within an international organization. We argued for a distinction between two types of agency in an international organization: first, the agency of governments as part of the norm-setting mechanisms of the organization, such as general assemblies and committees, and second, the distinct agency of the bureaucracy, or public administration, within the organization (Biermann et al. 2009: 39–40). We thus opened the black box of international organizations and focused on the internal bureaucracies and administrative bodies of intergovernmental organizations, with the aim to better identify and systematically study the autonomous agency of civil servants as *political agents* and as *policy entrepreneurs*.

*Managers of Global Change* also expanded the research focus from traditional international organizations, such as the World Bank or the International Maritime Organization, to the secretariats of international treaties. Especially in the field of global environmental politics, the number of treaties has tremendously grown over the last three decades, numbering now over 1,300. Most of these treaties have their own secretariat, and each secretariat has the potential to play an independent political role in the area that it covers. While some secretariats are tiny or integrated with existing UN organizations, others have grown into huge international bureaucracies, with hundreds of staff in new centers of global sustainability diplomacy, such as the former German capital of Bonn, which hosts around twenty secretariats. The secretariat to the UN climate convention, for instance, has evolved into a large international bureaucracy with around 500 employees and an annual budget of USD 90 million (Chapter 3).

The new focus on international public administrations, and its expansion to secretariats, did not only allow for a more nuanced understanding of international relations and for a more sophisticated empirical research program. It also helped to develop a new understanding of the political role and power of ordinary civil servants in often rather mundane technical agencies.

For example, our research has shown the discursive power of the secretariat of the UN desertification convention in preserving the concept of "desertification," which would have been less prominent if it were not for the discursive interventions of the secretariat's staff (Bauer 2009a). Our research also showed the discursive power of the economists in the Organization for Economic Co-operation and Development (Busch 2009b) and the powerful role of the civil servants in the tiny secretariat of the ozone treaties (Bauer 2009b). Our approach shed new light on the inner workings of international bureaucracies. For instance, we studied the professional backgrounds of civil servants in the International Maritime Organization (Campe 2009), dissecting their strong background in shipping, and in the World Bank with their unique culture shaped by traditional understandings of economics (Marschinski and Behrle 2009).

## 10.3 The New Contributions of *International Public Administrations* in *Environmental Governance*

These early studies of the 2000s, including *Managers of Global Change*, left many questions unanswered. The new ground charted in this earlier work needed more theoretical refinement, conceptual detail, and empirical data. This present volume is a milestone in driving this research agenda forward.

### *The Concept of "International Public Administrations"*

To start with, the conceptualization of "international public administrations" used in this book (Chapter 1) might be preferable to the term "international bureaucracies" used in *Managers of Global Change*. Both terms emphasize the important distinction between normative and administrative structures in international organizations, and the overlap between both terms is substantial (see, e.g., Wit et al. 2020). The term "international public administrations" might better link the study of national and international public administrations and more systematically merge national and international research into one fruitful research program (see also Bauer et al. 2017). The term "public administration" might also help shed earlier connotations of Weberian and more passive bureaucracies and open space for the more entrepreneurial and activist teams of international civil servants often seen in international political settings (e.g., Bauer 2009a; Siebenhüner 2009).

### *Conceptual Refinement*

Second, this volume offers more sophistication regarding the role of individual civil servants and the factors that determine their behavior. While *Managers of Global Change* had offered a set of variables under the heading of "people and procedures," the current volume goes a step further by adding more detailed conceptualizations of potential bureaucratic influence. An important innovation is the differentiation of administrative styles as informal behavioral routines of civil servants (Chapter 2; see also Bauer and Ege 2016). This focus on administrative styles, combined with a conceptualization of bureaucratic autonomy, allows for novel insights in the influence of international public administrations in global governance.

Similarly, Well et al. (Chapter 4) develop a convincing argument on a particular strategy that international public administrations use to increase their influence – "attention-seeking." This argument follows earlier claims that international public administrations reduce the information asymmetries in international negotiations by providing authoritative and more neutral insights on the issues at hand, especially for smaller countries with limited government capacities. However, as Well et al. show, to assume this position as a knowledge broker in negotiations, international bureaucracies first have to win the attention of negotiators. Attention-seeking thus becomes a central part of their strategic toolbox. Only by actively providing information to state representatives in international organizations can international bureaucracies insert their policy definitions and preferences in negotiations (Chapter 4; see also Jörgens et al. 2017).

In addition, this volume offers important insights on the role of the leadership of international public administrations, an issue that is notoriously difficult to analyze given the multiplicity of variables and the difficulties in designing comparative research designs. Hall (Chapter 5) takes on this challenge by carefully analyzing the role of the executive heads of the United Nations Development Programme, showing their vital impact on the expansion of the mandate of their organizations in times of shifting context conditions (see also Hall 2016).

In the end, however, this volume also shows that it is not free reign for international civil servants. One important constraint, as shown by Wagner and Chasek (Chapter 6), is still the budgetary control through governments, although even here international civil servants manage to keep some autonomy from powerful governments that tighten the purse strings. The financial control of governments illustrates the complex situation of international public administrations with universal membership but limited financial support: It is only the governments of the Global North that have the power to raise or cut funding and to use this influence over the policies of international bureaucracies, which increases the role of the major economies of the Global North and gives outsized powers to the citizens and voters in North America and Europe. This problem is well known for larger international organizations that suffered by the unilateral withholding of funding from some Global North countries, such as the United States. But Wagner and Chasek also highlight the smaller bureaucracies that are rarely seen in light of international financial dependencies. The Intergovernmental Panel on Climate Change, for example, depends for 95 percent of its income on only seventeen countries, with the United States alone contributing 39 percent (until 2017). The secretariat of the Intergovernmental Science-Policy Platform on Biodiversity and Ecosystem Services depends for 77 percent of its income on only four countries, Germany, Norway, the United Kingdom, and the United States (Chapter 6). While these funds are not conditioned on the outcome of these science assessments, one wonders what would happen if these assessments were to strongly counter the interest of those countries that pay for their secretariat. In the end, the "power of the purse," as Wagner and Chasek call it, stays the power of the Global North, counteracting the universal legislative assemblies of international organizations.

### *New Developments in Global Governance Theory*

Third, this volume connects theoretical insights on international public administrations with recent developments in global governance research. For example, new theoretical insights from orchestration research, developed over the last decade (Abbott and Snidal 2010; Abbott, Bernstein, and Janzwood 2020; Abbott et al. 2015), now help improve our understanding of the role of treaty secretariats, conceptualized by Hickmann et al. (Chapter 3) as "orchestrators" in global environmental governance. Through their orchestrating work, secretariats operate outside the traditional ground of intergovernmental diplomacy and the realm of foreign ministers and ambassadors in striped suits. As global orchestrators, international public administrations have become novel actors in multilevel governance settings, bringing in, and relying on, the energy and enthusiasm of civil society and local movements outside traditional state-led policymaking. Orchestration thus involves novel functions – such as citizen mobilization and partnership-building – that had not yet been part of the research design when we wrote *Managers of Global Change*.

This volume also brings in new normative considerations that had not been prominent in the early 2000s and in the *Managers of Global Change* program. One important question is the democratic legitimacy and accountability of international public administrations, which stands at the center of Bäckstrand and Kuyper's arguments (Chapter 8). Once international public administrations gain autonomous power and independent agency – and this book offers many examples for that – we need to interrogate the democratic quality of such bureaucracies, their leadership, and their internal decision-making. Given the unique context of global governance, however, we cannot simply transfer normative standards from national politics. Instead, as convincingly shown by Bäckstrand and Kuyper, we need to have different standards to hold international public administrations accountable. Participation, accountability and transparency, and deliberation are key elements for assessing the democratic legitimacy of acts of international public administrations and the degree to which those affected by such administrations have a say over these impacts on their lives (see Chapter 8).

Another normative standard, not prominent in this volume although present in many chapters, is the question of global equity and planetary justice (Biermann and Kalfagianni 2020). As with democratic legitimacy, also for global equity we need to ask how the autonomy and agency of international public administrations affect who gets what in global governance. A central concern is global distributive conflicts between the Global South and Global North, and especially the tension between the member assemblies of many organizations – often dominated by majorities of Global South countries – and the underlying funding structures that rely on a few "donor countries" from the Global North and that often draw only on voluntary contributions fluctuating year by year. The justice implications of the increasing autonomy and agency of international public administrations are an important research frontier still insufficiently covered by existing study programs.

## *Methodological Advancement*

Fourth, this volume provides ample evidence of the usefulness of new methods now available in the toolbox of the analyst. One approach, prominently represented in this book by Saerbeck et al. (Chapter 9), is social network analysis, building on a broader strand of work (e.g., Kolleck et al., 2017). Social network analysis allows us to gain a deeper understanding into the interdependencies and cooperative links among large numbers of international organizations and bureaucracies, in a way that grants new insights beyond what has been possible with the earlier case studies on small-*n* interlinkages. Social network analysis also allows to bring in large data-collection tools, such as Twitter analysis and, in this volume (Chapter 9), the generation of large datasets through surveys. Such approaches also allow for new theoretical understanding and conceptualization – for instance, the notion of an international or transnational administrative space that can be studied through such large-*n* approaches.

## *New Empirical Developments*

Finally, in addition to conceptual advancement and refinement, the studies in this volume present a vast array of fascinating new empirical developments. One example is Hickmann et al.'s study (Chapter 3) on the secretariat of the climate convention, directly relating to the earlier study by Busch (2009a) on the same topic. While Busch concluded in 2009 that the climate secretariat would "live in a straitjacket," not being able to develop its own policy agenda given strong pressures of governments in a highly conflictual policy field, Hickmann et al. now show that times have changed. In the wake of the 2009 Conference of the Parties in Copenhagen, widely seen as a disaster, the climate secretariat has worked itself out of their straitjacket, with the permission of governments that had lost collective leadership.

Michaelowa and Michaelowa (Chapter 7) add another perspective on the changing role of the climate secretariat, drawing on a large dataset that shows how the Clean Development Mechanism has influenced, and been influenced by, the climate secretariat. Their study might also give a glimpse of a future role of international public administrations in other domains with large financial transactions, for example, when it comes to global programs on carbon removal. The empirical example of the climate secretariat illustrates that in the realm of international public administrations, change in administrative policies, styles, and approaches is not only possible, it might even be more ubiquitous than expected. The example again shows the strong autonomous role of entrepreneurial staff of such international bureaucracies, which often is still neglected in more structural approaches to the study of international politics.

## 10.4 New Challenges: International Bureaucracies in the Anthropocene

Fifteen years after *Managers of Global Change*, it is time to reflect on the many changes that we have seen since then – conceptual changes that require a fresh look at global environmental politics but also broader political transitions that reshape our understanding of international organizations and bureaucracies.

### *International Public Administrations in the Anthropocene*

When *Managers of Global Change* was conceived as a research program around the turn of the millennium, the debates in the social sciences were still entrenched in the "environmental policy" paradigm. When writing *Managers of Global Change*, we did not hesitate to describe our unit of analysis as international "environmental" bureaucracies.

Today, such a perspective seems outdated, and many study programs have shown the deep interconnectivities between sectors that were earlier viewed as being distinctly environmental, economic, or social. The integration, or "nexus," between such sectors has become the focus of attention, along with a new understanding of coupled socioecological systems from local to planetary levels. Key challenges of our time, such as global heating or the massive loss of biodiversity, cannot be analyzed as environmental problems. Conversely, issues that were earlier defined as economic or social – such as poverty or inequality – are as much related to the exploitation of nature as to the exploitation of people.

The unique and novel planetary entanglement of people and nature is often described as the emergence of the *Anthropocene*, the geological age of humankind. Even though this term has been criticized because of apolitical "we-are-all-one-humankind" connotations (Biermann and Lövbrand 2019), all alternatives, such as "Capitalocene" (Moore 2017), have failed to catch on in the wider debate, and the neologism *Anthropocene* prevails. This new context of the Anthropocene invites us to adopt a new perspective on politics – and hence a new perspective on international public administrations. The traditional "environmental policy" paradigm has lost its luster (Biermann 2021), and today's "managers of global change" must bring a more complex and system-oriented perspective that goes beyond the "environmental managers" of the 1990s.

### *A World Environment Organization in the Post-environmental Age?*

This conceptual turn also raises the question of whether the long-standing call for the creation of a "world environment organization" still fits the needs of our time. This debate dates to the 1972 Stockholm Conference on the Human Environment, when first observers argued for the creation of an international agency for environmental protection (Bauer and Biermann 2005). In 1972, governments responded by establishing not a world environment organization but a less transformative UN program, the United Nations Environment Programme (UNEP), which is based since then in Nairobi, Kenya.

UNEP was never meant to be big and powerful. Its function was to serve as a catalysator and environmental conscience among the other agencies. Consequently, the secretariat of UNEP was designed to be small. Many elements typical for strong international organizations were withheld from UNEP: It lacks an operational mandate; its funding is voluntary and not based on assessed fixed contributions by governments; and the program has no formal right to initiate new international legal norms. Given these shortcomings, the debate for an "upgrade" of UNEP is as old as the program itself. Many scholars have called for the establishment of a full-fledged international organization on environmental protection, such as a United Nations Environment Organization or World Environment Organization. When I analyzed this debate over twenty years ago, I identified different ideal-types of such a world environment organization, from a hierarchical model with far-reaching powers to a less demanding cooperative model, and added an own proposal for a hybrid form of a world environment organization that I believed would significantly strengthen global environmental politics (Biermann 2000).

This lively policy debate found its culmination at the 2012 United Nations Conference on Sustainable Development in Rio de Janeiro, Brazil. While the European Union and the African Union with a few other countries called for an "upgrade" of UNEP, other countries objected, not the least the United States (Biermann 2013). In the end, no new agency was agreed, even though incremental reforms continued to strengthen UNEP over time. For example, a new United Nations Environment Assembly replaced the former governing council of UNEP and assumed some of the functions that proponents had envisaged for a world environment organization. New international regimes are now initiated by the United Nations Environment Assembly, mimicking the legislative functions of the International Labour Organization or the International Maritime Organization. And yet, the financial means of UNEP remain small and its financial base uncertain. Important debates and policy processes continue to develop outside the purview of UNEP, which has not much increased its standing as a global voice for the protection of key earth system processes.

In short, the incremental strengthening of UNEP, ongoing since the 1990s, remains important, and further steps in that direction are needed. In addition, however, the question arises whether other types of functional differentiation are needed to account for the complex interlinkages and nexus areas in global sustainability governance and the raising global inequalities between the North and the South. Here lies a major area for further research on international organizations and on the functioning of international public administrations in "earth system" governance (Biermann 2014).

## *Global Power Conflicts and Structural Injustice*

Regarding global power relations and conflicts, *Managers of Global Change* merely touched upon one key function of international bureaucracies that requires more systematic research and debate: the unique role of some international bureaucracies in supporting the interests of countries of the Global South in complex and often highly technical areas. Despite the autonomous agency of international bureaucracies, these bodies are still governed by intergovernmental assemblies, and most of these assemblies have voting majorities of developing countries that outnumber traditional powers in North America and Europe. Most UN organizations follow the principle of sovereign equality that grants each country one vote, regardless of its population size – and regardless of its economic or military might.

And yet, the power of developing countries in these assemblies is still limited. Most organizations depend on financial contributions of rich industrialized countries, prioritizing the "power of the purse" (Chapter 6); some organizations, such as the World Bank, even have special decision-making systems that prioritize industrialized countries. There is also a growing emphasis on alternative settings more open to Global North interests, such as the Organisation for Economic Co-operation and Development, the Group of 7 major economies, public–private partnerships and alliances, or informal settings such as the World Economic Forum.

In this situation, the often-large bureaucracies of international organizations, with their technical skills and expertise, can become important allies of smaller developing countries in helping them to raise their voice on complex issues. This grants – as we noted in *Managers of Global Change* – civil servants in such organizations new sources of authority. As one bureaucrat of the secretariat of the biodiversity convention noted, "As a national delegate it was my highest ambition to change at least one word in the text of the decision, as part of the secretariat I can influence the entire text" (cited in Siebenhüner 2009: 272).

## *New Anthropocene Challenges*

Finally, the Anthropocene has brought entirely new challenges for global governance and international cooperation. We need to ask whether today's international organizations and their bureaucracies are still apt to serve as "managers of global change" in increasingly dynamic, complex, and challenging policy environments.

One prominent example is global climate governance, which cuts across most traditional policies. Keeping global heating to less than 1.5°C will require huge investments in technology development and technology transfer, with a strong role for international public administrations to ease such knowledge and technology exchange. Global adaptation to a warmer world calls for international cooperation at unprecedent levels as well. International bureaucracies will need to engage more and in novel ways, for instance, when it comes to climate-related migration or the global provision of food. Moreover, most pathways that see the world staying within the 1.5°C warming scenario assume large-scale programs for carbon removal in the future, with techniques ranging from bioenergy with carbon sequestration and storage to the deployment of novel industrial processes for direct air capture. All these speculative approaches would require, if ever implemented, not only novel technologies but also new global governance mechanisms, from accounting systems for carbon removal to mechanisms that ensure global justice, food security, and global technology transfer. International organizations with strong bureaucracies would need to manage these novel types of global cooperation. International governance must also address the many other areas affected by climate change, for example, water shortages, sea level rise, or pressures on fertile land and food security caused by plant-based replacements of fossil fuels. And climate change is not the only area with such unique novel challenges for international organizations and bureaucracies. The Covid-19 pandemic, notably, has put new emphasis on the global health interdependencies and the importance of the World Health Organization in managing such crises; and there are many other global governance domains of growing global complexity and interconnectedness.

In short, while global interdependence is growing rapidly, the system of international organizations is still fixated on a model of diplomacy and cooperation that has not changed much since the twentieth century. The Charter of the United Nations was signed in 1945, and most international organizations have been created around that time. This volume makes an important contribution to a more nuanced understanding of the autonomous functioning of international public administrations; it lays vital groundwork for a renewed debate on how to transform international public administrations to more effectively address the multiple complex challenges of our century. And yet the book also shows how urgently we need novel, transformative models for effective and just international public administrations to cope with the pressing challenges of the twenty-first century.

## References

Abbott, K. W. and Snidal, D. (2010). International Regulation without International Government: Improving IO Performance through Orchestration, *Review of International Organizations* 5 (3): 315–344.

Abbott, K. W., Bernstein, S., and Janzwood, A. (2020). Orchestration. In F. Biermann and R. E. Kim (eds.), *Architectures of Earth System Governance: Institutional Complexity and Structural Transformation*, Cambridge: Cambridge University Press, 233–253.

Abbott, K. W., Genschel, P., Snidal, D., and Zangl, B. (eds.) (2015). *International Organizations as Orchestrators*. Cambridge: Cambridge University Press.

Barnett, M. and Finnemore, M. (2004). *Rules for the World: International Organizations in Global Politics*, Ithaca, NY: Cornell University Press.

Bauer, M. W. and Ege, J. (2016). Bureaucratic Autonomy of International Organizations' Secretariats, *Journal of European Public Policy* 23 (7): 1019–1037.

Bauer, M. W., Knill, C., and Eckhard, S. (eds.) (2017). *International Bureaucracy: Challenges and Lessons for Public Administration Research*. Basingstoke: Palgrave Macmillan.

Bauer, M. W., Eckhard, S., Ege, J., and Knill, C. (2017). A Public Administration Perspective on International Organizations. In M. W. Bauer, C. Knill, and S. Eckhard (eds.), *International Bureaucracy: Challenges and Lessons for Public Administration Research*, Basingstoke: Palgrave Macmillan, 1–12.

Bauer, S. (2009a). The Desertification Secretariat: A Castle Made of Sand. In F. Biermann and B. Siebenhüner (eds.), *Managers of Global Change: The Influence of International Environmental Bureaucracies*, Cambridge, MA: MIT Press, 293–317.

Bauer, S. (2009b). The Ozone Secretariat: The Good Shepherd of Ozone Politics. In F. Biermann and B. Siebenhüner (eds.), *Managers of Global Change: The Influence of International Environmental Bureaucracies*, Cambridge, MA: MIT Press, 225–244.

Bauer, S. and Biermann, F. (2005). The Debate on a World Environment Organization: An Introduction. In F. Biermann and S. Bauer (eds.), *A World Environment Organization: Solution or Threat for Effective International Environmental Governance?*, Aldershot: Ashgate, 1–23.

Bauer, S., Biermann, F., Dingwerth, K., and Siebenhüner, B. (2009). Understanding International Bureaucracies: Taking Stock. In F. Biermann and B. Siebenhüner (eds.), *Managers of Global Change: The Influence of International Environmental Bureaucracies*, Cambridge, MA: MIT Press, 15–36.

Biermann, F. (2000). The Case for a World Environment Organization, *Environment* 42 (9): 22–31.
Biermann, F. (2013). Curtain Down and Nothing Settled: Global Sustainability Governance after the "Rio+20" Earth Summit, *Environment and Planning C: Government and Policy* 31 (6): 1099–1114.
Biermann, F. (2014). *Earth System Governance: World Politics in the Anthropocene*, Cambridge, MA: MIT Press.
Biermann, F. (2021). The Future of "Environmental" Policy in the Anthropocene: Time for a Paradigm Shift, *Environmental Politics* 30 (1–2): 61–80.
Biermann, F. and Kalfagianni, A. (2020). Planetary Justice: A Research Framework, *Earth System Governance* 6: 100049.
Biermann, F. and Lövbrand, E. (eds.) (2019). *Anthropocene Encounters: New Directions in Green Political Thinking*, Cambridge: Cambridge University Press.
Biermann, F. and Siebenhüner, B. (eds.) (2009). *Managers of Global Change: The Influence of International Environmental Bureaucracies*, Cambridge, MA: MIT Press.
Biermann, F., Siebenhüner, B., Bauer, S., et al. (2009). Studying the Influence of International Bureaucracies: A Conceptual Framework. In F. Biermann and B. Siebenhüner (eds.), *Managers of Global Change: The Influence of International Environmental Bureaucracies*, Cambridge, MA: MIT Press, 37–74.
Bolton, J. (1994). *Speech in February*. Republished at www.c-span.org/video/?186235-1/un-ambassador-confirmation-hearing-day-1 (at 2 hours 48 minutes).
Busch, P.-O. (2009a). The Climate Secretariat: Making a Living in a Straitjacket. In F. Biermann and B. Siebenhüner (eds.), *Managers of Global Change: The Influence of International Environmental Bureaucracies*, Cambridge, MA: MIT Press, 245–264.
Busch, P.-O. (2009b). The OECD Environment Directorate: The Art of Persuasion and Its Limitations. In F. Biermann and B. Siebenhüner (eds.), *Managers of Global Change: The Influence of International Environmental Bureaucracies*, Cambridge, MA: MIT Press, 75–99.
Campe, S. (2009). The Secretariat of the International Maritime Organization: A Tanker for the Tankers. In F. Biermann and B. Siebenhüner (eds.), *Managers of Global Change: The Influence of International Environmental Bureaucracies*, Cambridge, MA: MIT Press, 143–168.
Hall, N. (2016). *Displacement, Development, and Climate Change: International Organizations Moving beyond Their Mandates*, New York: Routledge.
Hawkins, D. G., Lake, D. A., Nielson, D. L., and Tierney, M. J. (eds.) (2006). *Delegation and Agency in International Organizations*, Cambridge: Cambridge University Press.
Johnson, T. and Urpelainen, J. (2014). International Bureaucrats and the Formation of Intergovernmental Organizations: Institutional Design Discretion Sweetens the Pot, *International Organization* 68 (1): 177–209.
Jörgens, H., Kolleck, N., Saerbeck, B., and Well, M. (2017). Orchestrating (Bio-)Diversity: The Secretariat of the Convention on Biological Diversity as an Attention-Seeking Bureaucracy. In M. W. Bauer, C. Knill, and S. Eckhard (eds.), *International Bureaucracy: Challenges and Lessons for Public Administration Research*, Basingstoke: Palgrave Macmillan, 73–95.
Kolleck, N., Well, M, Sperzel, S., and Jörgens, H. (2017). The Power of Social Networks. How the UNFCCC Secretariat Creates Momentum for Climate Education, *Global Environmental Politics* 17 (4): 106–126.
Marschinski, R. and Behrle, S. (2009). The World Bank: Making the Business Case for the Environment. In F. Biermann and B. Siebenhüner (eds.), *Managers of Global Change: The Influence of International Environmental Bureaucracies*, Cambridge, MA: MIT Press, 101–142.

Moore, J. W. (2017). The Capitalocene, Part I: On the Nature and Origins of Our Ecological Crisis, *The Journal of Peasant Studies* 44 (3): 594–630.

Siebenhüner, B. (2009). The Biodiversity Secretariat: Lean Shark in Troubled Water. In F. Biermann and B. Siebenhüner (eds.), *Managers of Global Change: The Influence of International Environmental Bureaucracies*, Cambridge, MA: MIT Press, 265–291.

Wit, D. de, Ostovar, A. L., Bauer, S., and Jinnah, S. (2020). International Bureaucracies. In F. Biermann and R. E. Kim (eds.), *Architectures of Earth System Governance: Institutional Complexity and Structural Transformation*, Cambridge: Cambridge University Press, 57–74.

# Index

accountability, 149, 181, 185, 190, 193, 198, 232
Ad Melkert, 117
Adaptation Fund, 124
administrative styles, 5, 28, 32, 231
    measurement, 35
    operationalization, 35, 37
African Adaptation Programme, 120
agenda-setting, 79
agenda-setting bureaucracies, 90
An Inconvenient Truth (movie), 118
Anthropocene, 228, 235, 237
assessed contributions, 134, 140
attention-seeking bureaucracies, 12, 68, 73, 78, 86, 92, 99, 231

brokerage, 13, 74, 83, 231
budget, 8, 17, 62, 110, 132, 138, 156, 163
    negotiations, 18, 137, 138, 149
bureaucratic authority, 3, 74, 80, 203
bureaucratic autonomy, 31, 75, 157, 231. *See also* international public administrations (IPA)
bureaucratic reputation, 3
bureaucratic utility, 158

CBD, 7, 131, 140, 142, 209
CBD Secretariat, 7, 75, 85, 90, 96, 212, 237
Clark, Helen, 107, 121, 126, 127
Clean Development Mechanism, 11, 17, 124, 155, 164, 172, 234
    baseline and monitoring methodologies, 167
    standardized baselines, 169
climate adaptation, 107, 112, 120, 122, 125, 127
climate refugees, 127
collective principals, 10
conferences of the parties (COP), 7, 137, 190
Council of Europe, 7

delegation, 154
deliberation, 181, 186, 196, 198, 232
Derviş, Kemal, 116, 121, 126, 127
development organizations, 107
Draper, William, 113, 125, 127

earmarked funding, 135, 141
emanations, 10
Environment and Energy Group, 115
epistemic communities, 16
European administrative space, 203
European Commission, 16
executive heads, 93, 95, 108, 110, 127
Expanded Programme for Technical Assistance, 113
expertise, 8, 74, 76, 81, 157, 167, 203
external resources, 166

FAO, 27, 43, 45
framing, 139

G7/8, 125
G20 agenda, 125
GEF, 113, 114, 145
gender equality, 108
global administrative space, 15, 16, 19, 201, 203, 207, 208, 215, 219, 233
Global Climate Action, 180, 183, 198
global climate governance, 12, 14, 57, 68, 83, 94, 202, 207, 234, 237
global environmental administrative space, 15, 202, 207, 215
global environmental governance, 202
good governance, 116
Green Climate Fund, 124

health, 108
High-Level Champions, 198
Human Development Report, 118
human rights, 108
humanitarian organizations, 107

ILO, 43, 45, 236
IMF, 43, 46
IMO, 230, 236
information, 162
institutional fragmentation, 7, 12

241

international bureaucracies, 154, 162, 203, 205, 228, 229. *See also* international public administrations (IPA)
international carbon markets, 161
international civil servants, 3, 4, 15, 31, 93, 111, 132, 154, 182, 229, 230, 237
 motivation, 78
international climate negotiations, 62
*International Cooperative Initiatives*, 64
international environmental bureaucracies, 1, 9, 12, 18, 27, 29, 57, 108, 219, 220, 228, 234
international environmental negotiations, 8, 17, 59, 121, 131
international environmental secretariats, 1, 74, 84, 131, 161, 201, 215, 223
international organizations, 2, 3, 76, 155, 160, 180, 182, 215, 219, 220, 228, 234, 236
 mandates, 107, 109
international public administrations (IPA), 1, 5, 27, 42, 73, 99, 131, 132, 153, 180, 201, 202, 204, 206, 208, 212, 220, 228, 230, 234, 236
 as actors, 1, 2, 4, 10, 13, 18, 59
 administrative styles, 6, 28, 32, 40
 as agenda-setters, 74
 autonomy of, 5, 9, 12, 59, 110, 233, 238
 core budget, 138, 140
 entrepreneurial IPAs, 29, 37, 44, 87, 234
 financing, 8, 17, 32, 65, 111, 120, 121, 124, 132, 139, 149, 163
 formal autonomy, 9, 28, 40
 influence of, 6, 15, 27, 30, 47, 58, 73, 80, 81, 134, 153, 162, 202, 231
 mandate expansion, 10, 108, 110, 124, 125, 127
 mandates, 8, 10, 61, 78, 86, 126, 203
 preferences, 31
 resources, 155
 servant-oriented IPAs, 28, 36, 45
 sources of influence, 8, 13, 41, 47, 57, 59, 60, 66–68, 159, 162, 166, 167, 172
 staff, 16, 17, 31, 77, 82, 87, 111, 115, 120, 137, 160, 163, 171, 172
 structural autonomy. *See* international public administrations (IPA):formal autonomy
international secretariats, 78, 84, 153, 180, 206, 229
IOM, 43, 127
IPBES, 18, 131, 136, 145, 232
IPCC, 18, 118, 131, 136, 144, 211, 232
issue-linkage, 82, 89, 96

Ki-Moon, Ban, 125
Kyoto Protocol, 11, 89, 124, 164

leadership, 4, 81, 95, 111
Least Developed Countries Fund, 123, 124
legitimacy, 8, 75, 81, 94, 181
 democratic, 12, 181, 184, 190, 197, 232
Lima-Paris Action Agenda, 13, 58, 66, 183, 190

Malloch-Brown, Mark, 115, 126, 127
*Managers of Global Change*, 1, 49, 228, 234
Marrakech Partnership for Global Climate Action, 12, 183, 190
methodological approaches, 7, 17
Millennium Development Goals, 122
Millennium Development Goals Carbon Facility, 117
mitigation, 115
Momentum for Change initiative, 58, 64, 94, 142, 150, 189
Montreal Protocol, 124
multilateral negotiations, 6, 11, 18, 74, 84, 99, 189, 202
 role of chairpersons, 11, 78, 79, 81, 91
multilateral treaty-making, 60
multilevel governance, 206, 219, 232
multiple principals, 10

National Adaptation Programmes of Action (NAPA), 124
network centrality, 8, 13, 14, 32, 84, 92, 209, 220
Non-state Actor Zone for Climate Action, 13, 58, 67, 183, 190
non-state actors, 57, 63–65, 98, 111, 136, 180, 183, 189, 194, 201, 205, 215

OECD, 7, 27, 30, 43, 45, 230
Office of the UN High Commissioner for Human Rights, 127
orchestration, 12, 13, 57, 60, 67, 79, 180, 182, 185, 190, 197, 232
Organization for the Prohibition of Chemical Weapons (OPCW), 109
organizational autonomy, 3, 9, 12
organizational culture, 166
organizational design, 4
organizational development, 9
organizational environment, 8
organizational expansion, 124
organizational routines, 28
organizational structure, 4
OSCE, 43, 46
overlap management, 7, 15, 96, 112. *See also* institutional fragmentation

Paris Agreement, 11, 12, 14, 17, 63, 67, 75, 86, 93, 108, 160, 188, 207
participant observation, 14
participation, 180, 185, 190, 197, 232
planetary justice, 233
policy cycle, 74
policy networks, 78, 82, 92, 96, 183, 202, 207–209, 211
policy preferences, 8, 11–13, 16, 17
poverty reduction, 116
principal–agent relationship, 2, 10, 28, 61, 73, 75, 110, 132, 133, 149, 154, 158, 180, 182, 223
programmes and budgets, 132

resource dependency theory, 111
Responsibility to Protect, 6
Rio Conference, 113, 160, 235

secretariat leadership, 6, 11, 17, 31, 45, 84, 93, 94, 138, 143, 231
social network analysis (SNA), 7, 14, 202, 208, 233
   inferential techniques of, 14
   survey-based, 15
sociological institutionalism, 3
Special Climate Change Fund, 123
Speth, James Gustave, 113, 114, 127
stakeholders, 19, 63, 67, 93, 98, 192
Stern Review on the Economics of Climate Change, 118, 125
Stockholm Conference, 113, 235
sub-national actors, 57, 63, 66
surplus, 164
sustainable development, 113, 122

transgovernmental networks, 206
transnational institutions, 60
transparency, 181, 185, 193, 198, 232
trust funds, 142

UN Environment Assembly, 236
UN General Assembly, 113, 135
UN secretariat, 5, 125
UN Secretary-General, 6, 62, 66, 180, 190
UN Special Fund, 113
UNCCD, 131, 137, 230

UNCCD Secretariat, 7, 90
UNCED, 113, 114, 125
UNCTAD, 6, 62
UNDP, 108, 113, 231
   Climate Change Strategy, 119
UNEP, 110, 155, 235
UNESCO, 7, 27, 43, 45
UNFCCC, 11, 62, 86, 92, 108, 121, 131, 142, 155, 160, 180, 207, 208
   Conference of the Parties, 63
UNFCCC Secretariat, 11, 57, 61, 62, 66, 67, 75, 85–87, 92, 142, 155, 160, 172, 181, 183, 188–190, 197, 229, 233
   mandate, 61, 63, 64
UNHCR, 43, 44, 110
UNICEF, 113
UNIDO, 120
UN-WOMEN, 127

visibility, 11
voluntary contributions, 135, 140

WHO. *See* World Health Organization (WHO)
World Bank, 109, 110, 230, 236
World Bank Environment Department, 7
World Commission on Environment and Development, 113
world environment organization, 235
World Food Programme (WFP), 120
World Health Organization (WHO), 43, 113, 237
World Resources Institute, 114